Disaster Recovery Through the Lens of Justice

"This book fills an important gap in the field of public administration and emergency management by providing both academics and practitioners with a wealth of perspectives linking complex issues of community resilience to recovery. In many recent disasters there has been a marked difference in the level of recovery of local, sometimes adjacent, communities. Some communities have recovered with almost no lingering impacts, while other communities, often those consisting of marginalized populations, struggle to attain normalcy years later. This book places special emphasis on vulnerability and social justice during long-term recovery. Throughout the book, clear principles for guiding policy-makers and administrators are presented, providing a very useful framework for answering the difficult question of 'resilience for whom?'"
—John Kiefer, *Professor, Director, Master of Public Administration Program, University of New Orleans, USA*

Alessandra Jerolleman

Disaster Recovery Through the Lens of Justice

Alessandra Jerolleman
Jacksonville State University
Metairie, LA, USA

ISBN 978-3-030-04794-8   ISBN 978-3-030-04795-5 (eBook)
https://doi.org/10.1007/978-3-030-04795-5

Library of Congress Control Number: 2018964984

© The Editor(s) (if applicable) and The Author(s), under exclusive licence to Springer Nature Switzerland AG 2019
This work is subject to copyright. All rights are solely and exclusively licensed by the Publisher, whether the whole or part of the material is concerned, specifically the rights of translation, reprinting, reuse of illustrations, recitation, broadcasting, reproduction on microfilms or in any other physical way, and transmission or information storage and retrieval, electronic adaptation, computer software, or by similar or dissimilar methodology now known or hereafter developed.
The use of general descriptive names, registered names, trademarks, service marks, etc. in this publication does not imply, even in the absence of a specific statement, that such names are exempt from the relevant protective laws and regulations and therefore free for general use.
The publisher, the authors, and the editors are safe to assume that the advice and information in this book are believed to be true and accurate at the date of publication. Neither the publisher nor the authors or the editors give a warranty, express or implied, with respect to the material contained herein or for any errors or omissions that may have been made. The publisher remains neutral with regard to jurisdictional claims in published maps and institutional affiliations.

Cover illustration © Melisa Hasan

This Palgrave Pivot imprint is published by the registered company Springer Nature Switzerland AG
The registered company address is: Gewerbestrasse 11, 6330 Cham, Switzerland

Acknowledgments

I have to start by thanking my amazing daughter, Catalina, who has spent much of her young life listening to me talk about injustice and willingly shared my time with this manuscript and other work. I would also like to thank Liliana Colareta for her constant help and support, both with Catalina and otherwise, without which this book would not be possible.

Special thanks go to Leila Darwish and Paula Buchanan, both of whom provided assistance with the research for this book, including the writing of small vignettes, as well as to the long list of colleagues who have allowed me to bounce ideas off them and greatly enriched my thinking and analysis.

Contents

1 **Introduction: Recovery, Resilience, Vulnerability, and Justice**   1
   1.1 *Applying a Justice Paradigm*   4
   1.2 *Conceptualizing Justice*   5
   1.3 *Disaster Vulnerability and Resilience*   10
   1.4 *Deserving Victims and Post-Disaster Fraud*   11
   1.5 *Survivor Agency*   13
   1.6 *Social Capital and Collective Efficacy*   14
   1.7 *Public Policy and Legislation*   16
   1.8 *Implementation Challenges*   17
   1.9 *Disaster Risk Reduction and Creation*   18
   1.10 *Disparate Outcomes*   18
   1.11 *Resilience for Whom?*   19
   1.12 *Defining Just Recovery*   20
   References   22

2 **Deserving Victims and Post-Disaster Fraud**   27
   2.1 *Deserving Victims*   29
   2.2 *Duplication of Benefits and Appeals*   30
   2.3 *Deserving Victims, Fraud, and the Corrosive Community*   32
   2.4 *Deserving Corporations*   35
   2.5 *Government Corruption*   40
   2.6 *Deserving Victims, Post-Disaster Fraud, and Justice*   40
   References   41

3 Survivor Agency   45
3.1   Survivor Agency   46
3.2   Collective Efficacy and Social Capital   49
3.3   Reclaiming Agency   51
3.4   Survivor Agency and Justice   52
References   53

4 Public Policy and Legislation   55
4.1   US Disaster Policy   56
4.2   Federalism   57
4.3   Federal Disaster Programs   60
4.4   Human Rights and Disaster Recovery   63
4.5   Public Policy, Legislation, and Justice   64
References   65

5 Implementation Challenges   67
5.1   Program Implementation   68
5.2   Implementation of Existing Policies and Programs   70
5.3   Limitations of Current Programs and Policies   71
5.4   Using Current Policies and Regulations to Promote Just Outcomes   75
5.5   Implementation Challenges and Justice   75
References   76

6 Disaster Risk Reduction and Creation   79
6.1   Disaster Risk Creation   80
6.2   The Safety Premium   82
6.3   Considering Justice in Resilience   84
References   85

7 Disparate Outcomes   87
7.1   Poverty and Economic Hardship   88
7.2   Impacts to Public Housing   90
7.3   Renters   91
7.4   Displacement   92
7.5   Just Recovery   93
References   95

8	Conclusion: Resilience for Whom?	97
	8.1 *Revisiting Just Recovery*	98
	8.2 *Person and Community-Centered Recovery*	100
	References	101

Index 103

List of Boxes

Box 2.1　Deserving Victims and Federal Assistance: The Case of Puerto Rico　34
Box 2.2　Deserving Corporations: The Case of Puerto Rico　38
Box 4.1　Public Assistance for Infrastructure　61

CHAPTER 1

Introduction: Recovery, Resilience, Vulnerability, and Justice

Abstract Disparate outcomes from disaster, including the permanent displacement and essentially forced internal migration of lower-income and minority residents, occurred following Hurricane Katrina in 2005 and continue to occur following large disasters as well as chronic events. The introduction uses examples such as Hurricane Katrina to describe the nexus between justice and disaster recovery outcomes, arguing for the application of a capacities justice framework and a human rights approach in disaster recovery. The chapter provides some background on the existing literature and policy frameworks, including the concepts of resilience and vulnerability, along with the role of historic practices and systemic racism in the creation of the modern hazard landscape.

Keywords Disaster justice • Capacities justice • Disaster recovery • Climate adaptation • Disaster resilience • Disaster vulnerability

Over the five years following Hurricane Sandy, houses along the New Jersey coastline have become larger and taller than they had been prior to the hurricane. Median home prices have risen, while the percentage of full-time households, as opposed to second homes, has declined. As with many coastal communities across the United States, some measure of population shift and even coastal gentrification was already under way prior to the storm, but as often happens, the combination of damage to older structures and the inability of residents living on fixed incomes to fully recover accelerated the trend (Terruso 2017).

Similarly, disparate outcomes, including the permanent displacement and internal migration[1] of lower-income and minority residents, occurred following Hurricane Katrina in 2005 and continue to occur following large disasters as well as chronic events. In 2017 alone, over one million Americans were displaced from their homes by disasters (Goodell 2018). These outcomes are not solely a result of the disaster[2] itself but in fact largely stem from a history of disparities in health and well-being that serve to concentrate vulnerability within minority, working class, and poor communities[3]—who are then not given full access to resources or sociopolitical decision-making processes in disaster recovery because their voices and their knowledge are not valued by technocratic recovery processes and their presence is considered undesirable by elites (Allen 2007). For many, the price of adaptation is too high and leaving becomes the only option,[4] while others cannot afford to migrate and remain behind in communities that are deteriorating and lack the resources to offer basic services (Goodell 2018).

Climate change is a global problem in which both the harms and the responsibility have been unequally distributed across time and geography. Despite benefiting the least from the global marketplace and development trends that have exacerbated climate impacts, vulnerable groups will be the most adversely impacted (Swim and Bloodhart 2018). Large-scale climate migration is expected in the future, with impacts primarily upon the poor, and it is already occurring in other nations and in areas of the United States. In fact, there is a long history of disaster-influenced migration within the United States, including impacts from the Dust Bowl, and massive African American migration following the Great Mississippi River Flood of 1927 (Rivera and Miller 2007).

Despite clear evidence of unjust outcomes, and of a system that cannot meet the needs of the entire impacted community, the 2017 hurricane season once again repeated and exacerbated these patterns of injustice. Hurricane Harvey, for example, a record-setting rainfall event for the Houston area, with over 55 inches of rainfall, impacted immigrant populations, minority households, and low-income families in disproportionate ways (Kelly et al. 2017). One study indicated that after Hurricane Harvey, Hispanic and black residents were twice as likely to experience an income shock as their white counterparts and routinely did not receive the assistance they needed (Maxwell 2018).

The direct impacts of Hurricane Harvey were significant, with over 25% of Harris County under water at peak flooding, over 13,000 persons rescued, and over 37,300 in shelters (Norton et al. 2018). The indirect, and cascading, impacts are proving to be even greater, as only 20% of impacted households carried flood insurance (FEMA 2018). One study found that as many as 30% of residents were behind in mortgage or rent payments as a result of disaster losses, or impacts to employment following Harvey, and that around 25% were struggling to pay for food (Goodell 2018).

According to several estimates, losses from natural[5] disasters in 2017 were the highest recorded in US history, exceeding $306 billion. Wildfires, alone, led to over $10 billion in damages in Napa, Sonoma, and Mendocino Counties in California (Mark 2018). The rising costs of disaster impacts, coupled with several projections of ever-worsening impacts, have been utilized by politicians and others to argue for greater local investment and responsibility for disaster risk reduction and recovery costs, including increased personal responsibility (Pew Charitable Trusts 2018). This narrative is further used to excuse failures in places such as Puerto Rico, which, as a US territory, lacks the political clout of Texas.

Hurricane Maria caused over $90 billion in damages to Puerto Rico, making it the third costliest cyclone in the United States since 1900. The resulting damage to the power grid, water infrastructure, healthcare, and other essential services led to a series of cascading impacts and an ongoing public health emergency. Puerto Ricans, on average, lacked cell service for 41 days, water for 68 days, and went 84 days without electricity (Palm Beach Post 2018). Although the initial death count was set at 64, subsequent studies have set the death count as high as 4645. As of August 2018, the official death count was revised to 2975 (Fink 2018). These deaths were a direct result of lack of water, a faltering power grid, healthcare disruption, and a general lack of essential services (Hernandez and McGinley 2018). Unlike Texas, Puerto Rico lacks any voting power in the US Congress, having only one delegate without an official vote (Palm Beach Post 2018).

This book draws upon an analysis of the official and media reports following Hurricanes Harvey and Maria to illustrate the fact that the challenges described in this text have continued through to more recent events. Examples from Hurricanes Maria and Harvey are included in each chapter, as well as older examples following Hurricane Katrina and other events, where more peer-reviewed research is available.

1.1 Applying a Justice Paradigm

As the preceding examples clearly indicate, there is a clear nexus between justice and disaster recovery outcomes, with an unequal distribution of disaster risk and vulnerability—creating a cycle in which pre-event vulnerabilities are exacerbated by disaster impacts and failure to fully recover is framed as a personal and community failing to be resilient. There has been some exploration by social scientists, and others, of unjust and unfair outcomes in disaster recovery, including a robust literature on vulnerability that dates back several decades. There is a general understanding within the literature that current post-disaster policies and frameworks disadvantage the most vulnerable and lead to disparate outcomes, primarily through the perpetuation of structural inequalities (Wisner et al. 2005; Pelling 2003).

However, the mechanisms by which this occurs, and the extent to which the regulations and agency cultures drive this outcome, are not fully understood. There is also a general understanding that social and economic structures, including land use policies and historic practices, such as redlining, have concentrated hazard risk into vulnerable zones whose inhabitants do not benefit from the very policies that create and increase their risk (Freudenburg et al. 2008; Tierney 2010). This text will synthesize existing research across multiple disciplines, including public administration, political science, sociology, anthropology, psychology, and economics, in order to improve our understanding of the complexity of interactions between these mechanisms in the context of disaster resilience. This synthesis is coupled with real-world examples from more recent events, where information is available.

To that end, this discussion will bring together existing literature, seeking to integrate the language and theories of justice more fully with the concepts of resilience and vulnerability, including the role of historic practices and systemic racism in the creation of the modern hazard landscape. Some researchers have applied concepts from environmental justice to the challenge of climate adaptation, but this literature remains limited to particular cases such as Hurricane Katrina. Even when a justice framing is applied, there is a lack of clarity regarding the conceptualization of justice being utilized, as described by Allen (2013) in her analysis of justice impacts on Non-Governmental Organizations (NGO) assistance and outcomes that utilized the framings of distributive justice, procedural justice, cultural justice, representative justice, and capacities justice. In fact, it is often the case that communities utilize multiple definitions and framings of justice in order to conceptualize their claims of harm and advocate for improvements. Although a strict definition may not be necessary in order for community leaders to advocate for a reduction in injustice, and at times the ability to use a wide range of terminology that responds to the particular political framing in use can be beneficial, it is still important to acknowledge that the definitions implicitly or explicitly embedded within strategies and techniques of NGOs and community organizations do lead to different results (Allen 2013). More often than not, justice itself is not explicitly framed, but overt examples of injustice are called out and identified. This type of approach can serve to bring attention to particular actions, but cannot in and of itself drive significant structural change. This book proposes a more purposeful definition of justice in order to create more concrete policy options for the reduction of injustice.

1.2 Conceptualizing Justice

The majority of efforts to apply a justice paradigm to the study of disasters and disaster risks have originated from the study of environmental justice. An environmental injustice occurs when an individual or a group bears disproportionate risks, has unequal access to environmental goods, or has less opportunity to participate in environmental decision-making (Shrader-Frechette 2002). This book argues that the same measure could readily be applied to the creation of risk in other contexts, such as along coastlines or within floodplains, where individuals or groups are asked to bear disproportionate risks without having proportionate access to the benefits of development or economic decisions that have exacerbated their risk. In other words, a social injustice is perpetuated through the creation and exacerbation of disaster risk. This book argues that there can be no resilience if there is no justice and that recovery cannot truly occur if it is not just. To this end, the book will utilize the phrase "Just Recovery"[6] to encompass all of the elements of a just approach to resilience and recovery. This term will be further defined later in this chapter.

The intricate connection between environmental injustice and disaster risk—through the degradation of environmental resources that protect communities; the siting of hazardous facilities in impoverished, indigenous, or minority communities; or other mechanisms that increase community vulnerability—makes this application of an environmental justice lens particularly apt. It is often the case that disasters stemming from natural events such as hurricanes combine with the built environment to create technological disasters, such as when the levees failed in Hurricane Katrina or when toxic materials are released as a result of a flood—often into the backyards of the poor communities in which the industrial facilities are located. Take, for example, the fact that of the 21,600 facilities that handle toxic materials across the United States, over 1400 can be considered at high risk of flooding. In just one incident, during Hurricane Harvey, a chemical plant in Baytown, Texas, released over 34,000 pounds of sodium hydroxide and 300 pounds of benzene, the majority of which entered the floodwaters (Adriano 2018).

Further illustration of the connection between environmental justice and disaster injustice is found in the utilitarian arguments made against environmental justice efforts, arguments that practically mirror those made regarding exposure to risk from hazards. These arguments often begin from the presupposition, or claim, that victims derive benefits from living in the vicinity of toxic facilities, or other dangerous sites, because of employment or other economic benefits. This is essentially the same logic that supposes that households living at risk in environmentally fragile areas are benefiting from their location or that they are complicit in the degradation through their participation in extractive industries—despite having no viable alternative livelihood. It is clear that the benefits of inhabiting risky areas, and that the economic benefits that accrue from the process of creating such risk, do not accrue to the inhabitants of the risky locations (Freudenburg et al. 2008). A clear example of this can be found in Appalachia, where absentee land ownership has led to a concentration of political and economic power in the hands of external interest, leaving smaller land owners with less bargaining power. The proceeds from coal production are funneled away from the community, while the negative impacts on housing, education, health, and so on, remain within the community (Shrader-Frechette 2002). Although inhabiting risky areas, such as the locations in the proximity of industrial facilities, or oil and gas exploration, can provide employment opportunities, these opportunities come with high health risks and consequences—and at the expense of other alternatives.

Another common argument is that the correlation between hazardous sites and minority populations is not proof of racism or injustice but is instead coincidental (Shrader-Frechette 2012). This argument is, at times, also made regarding the concentration of minority communities within floodplains or in riskier coastal areas such as those outside of protective levee systems. However, there is an extensive body of research indicating that social vulnerability, including exposure to risky environments, is not evenly distributed among social groups or between places (Cutter and Emrich 2006). This argument will be further discussed in Chap. 6 "Disaster Risk Reduction and Creation."

In order to fully apply this concept to the study of disasters, it is necessary to begin with an overview of the trajectory of theories of justice, both within the study of environmental justice and in other arenas. The framing of justice utilized by environmental justice scholars originated largely from a Rawlsian focus on fairness and distribution of goods (Allen 2013). The concept of distributive justice is based upon an equal apportioning of the social burdens and benefits but does not consider the role that institutional contexts can play in determining distribution. Essentially, environmental justice requires a more equitable distribution of goods and bads, along with more public participation in the process of distribution and evaluation (Shrader-Frechette 2002).

One example of this is land use and zoning, where both local and federal policies can serve either to promote injustice, as in the case of redlining by the federal government, or to support equitable siting. Federal policies can, at times, serve to control local vested interests, such as those which might seek to prevent particular communities from obtaining housing in more desirable areas, but they can also benefit corporate interests as is the case with the Coastal Zone Management Act, which has largely helped energy development along the coast while leaving residents with the financial burdens and risks of offshore oil extraction (Shrader-Frechette 2002).

Over time, environmental justice scholars moved beyond simply framing justice as equitable or fair distribution, focusing as well on the ability of people to participate in the decisions that affect them and their exposure to risks, termed procedural justice. Full and informed participation is considered integral to the concept of procedural justice, a concept seen by some as an alternative to distributive justice and by others as a necessary corollary. Full participation requires a complete understanding of the hazards imposed by a particular set of decisions. However, there are many obstacles that stand in the face of full understanding, including technocratic framings that privilege particular types of knowledge and legal protections such as non-disclosure. Additionally, a full understanding of the consequences may not be possible at the time of decision-making, as some longer-term impacts cannot be fully predicted. As a result, consent may be given without adequate information or, in some cases, with erroneous information being provided to current or potential residents of at-risk neighborhoods and with no opportunity to revisit their consent as more information is revealed. It is not uncommon to have government and industry experts providing reassurance in the face of community concerns and even conflicting data (Ottinger 2012).

One of the primary obstacles to procedural justice lies in the strong property rights culture within the United States that often privileges the rights of landowners at the expense of impacted neighbors. Achieving equal access to public goods, such as air quality or clean water, may require restricting property rights, including the rights to utilize natural resources. This is especially true for resources that are finite or non-renewable (Shrader-Frechette 2002). This obstacle will be further explored in Chap. 4—Public Policy and Legislation.

Allen (2013) provides an overview of the further expansion of justice framings to consider concepts such as cultural justice, social justice, representative justice, and eventually capabilities and capacities. All of these framings of justice are connected by Allen in a trajectory where procedural justice can be considered an offshoot, or subset, of distributive justice that links institutions to the distribution of goods. The distribution of goods, or maldistribution,[7] is then directly linked to social context, providing a link to the work of cultural justice scholars as well as to the literature on vulnerability, which will be discussed later in this introduction, with a further connection to representative and participatory framings of justice through its efforts to promote full partnership and participation (including equal right to self-determination in societal decision-making) within the social context and with institutions. The concept of full participation is an important lynchpin, as it requires resources and must therefore address maldistribution and inclusion. A capacities approach to justice sees the transformation of goods into a fully lived life as the true measure of justice, going beyond rules and procedures to actually look at the types of lives people are able to lead (Allen 2013).

This kind of approach, at times also referred to as a capabilities approach, argues that justice is primarily about a person's ability to live a life they consider worthwhile and is therefore comparative and not absolute. Under this conceptualization, justice depends on individual values and requires further exploration of what is meant by well-being, both hedonic (subjective well-being) and eudaimonic (not just happiness but also the ability to flourish and be satisfied with life). Edwards et al. (2016) seek to further the recognition of the capabilities approach to justice and argue that a shift has been occurring in recent literature although it is not yet prominent outside of psychology. A similar conceptualization of justice is offered by Zakour et al. (2018), who argue for a social justice framework for disasters, stemming from the recognition of the importance of values, beliefs, culture, customs, way of life, and so on. This framing of social justice, based upon adaptive capacity (the ability to meet human needs and maintain social structures without adverse impacts) and transformative capacity (the ability to fundamentally alter the socio-ecological system in order to reduce risks), is also aligned with the capabilities approach, positioning capabilities as the foundation of both adaptive and transformative capacity. This approach opens the door to the consideration of personal agency and community efficacy, both of which are discussed in Chap. 2—Deserving Victims and Post-Disaster Fraud.

Free and informed consent, a premise of environmental justice, is also critical to social justice in the face of disaster risks. According to Shrader-Frechette (2002), free and informed consent is only possible in cases when there is full disclosure of the threat, potential victims are competent to evaluate the threat, and the danger is both completely understood and voluntarily accepted without coercion. This understanding of consent presupposes that those upon whom threats are imposed have access to safety and are not manipulated into accepting risks due to harsh economic realities or held to a prior consent based upon insufficient information.

The concept of Just Recovery, applying a justice lens to the conceptualization of disaster resilience, begins with the fundamental assertion that survivor agency and community efficacy must be supported through ongoing disaster risk reduction processes, community planning, disaster recovery, and other processes that create and reduce disaster risks and impacts. This book utilizes the capabilities justice framework, as defined by Allen (2013), including procedural and distributive justice. Another important concept, drawn from the work of Shrader-Frechette (2002), is that of the principle of prima facie political equality (PPFPE). This principle is based upon the work of both John Rawls and Bruce Ackerman and presumes that any different or unequal treatment of others must be justified by the discriminator, as only equality is defensible. It is based upon three principles: political equality, economic equality, and equal opportunity. This concept of PPFPE is a value addition to the capabilities justice framework, as it directly places the burden of proof upon the discriminator and not upon the victims. Current policies do just the inverse, requiring victims to assert their rights to a livable environment.

1.3 Disaster Vulnerability and Resilience

A great deal has been written regarding both disaster vulnerability and disaster resilience, including the role of social capital in both. This book will not provide an extensive discussion of these concepts, beyond their application to the larger discussion of disaster justice. However, it is important to position this text within the larger discussion and to clarify what is meant by both disaster vulnerability and resilience within this book. Zakour and Swager (2018) have proposed a Vulnerability-Plus (V+) theory that aims to build upon both Ben Wisner's Pressure and Release Model (PAR), and the Access Model, both focused on vulnerability, and to blend these theories of vulnerability with the resilience literature, creating one framework.

The PAR model presents vulnerability as a result of the impacts of social processes over time, where root causes create dynamic pressures that are then translated into unsafe conditions. These root causes and dynamic pressures interact with hazards to create disaster risks. Zakour and Swager (2018) suggest that the Access to Resources model of vulnerability is more dynamic than the PAR model, with its explicit focus on the lack of capabilities leading to fragile livelihoods and unsafe locations. The Access model looks at five different types of capital: human, social, physical, financial, and natural, as well as the impacts of root causes, such as structures of domination and social relationships, on access to resources across these five capitals.

Similarly, theories of resilience have looked at resource types that align with the five capitals in many ways. These have included economic development, social capital, information/communication, and collective action. These resources vary in terms of robustness (strength and accessibility, ability to withstand stress), redundancy (ability to substitute for other system elements), and rapidity (speed of mobilization). These concepts of robustness, redundancy, and rapidity can also be applied to the five types of capital utilized within the Access model, allowing for a deeper understanding of the ways in which those capitals are accessed and utilized, as well as the ways in which disasters can impede the ability to access those capitals. Just Recovery empowers individual agency in support of collective action, allowing for the exercise of resilience in support of recovery.

The V+ theory emphasizes both the structural and economic variables that are commonly used within vulnerability theories, as well as the individual and collective outcomes employed by resilience theories. The combination of these frameworks allows for a richer analysis of the influence of both root causes and structural constraints on resilience and vulnerability, enriching the chain of causality that forms the foundation for the PAR. Combining these frameworks also allows for greater consideration of individual agency, something that is not explicitly taken into account by the vulnerability models (Zakour and Swager 2018). A truly resilient recovery must effectively mobilize community resources and capabilities, increasing access to resources and seeking wellness as the desired outcome. The consideration of wellness as a critical outcome is consistent with the concept of Just Recovery. Under this model, wellness is considered a necessary outcome of resilience and is defined as including lack of disease, adequate role functioning, lack of generalized distress, and high quality of life. This focus on wellness provides a clear linkage between the V+ theory and the theoretical framing of justice utilized by this book. The Vulnerability-Plus theory is based upon 12 assumptions listed fully within the endnotes.[8]

1.4 Deserving Victims and Post-Disaster Fraud

During and after Hurricane Katrina, poor households found themselves in a vicious cycle of having to spend their savings and sell productive assets, expending limited resources on coping actions that later undermined the long-run sustainability of the household. Households were often splintered, as individuals were forced to migrate for economic reasons or to ensure access to healthcare and education. Some household members were forced into dangerous, illegal, or culturally inappropriate livelihoods (Zakour et al. 2018). These challenges were being faced while the broader media and cultural narratives portrayed victims of Hurricane Katrina as undeserving, placing the impetus for recovery and the blame for their misfortune entirely upon their shoulders. Such a narrative ignores the role that structural violence plays in the perpetuation of exclusion and inequity through socio-political institutions and policies—such as those tied to disaster recovery (Zakour et al. 2018).

The dominant narrative in the United States is one of equality, justice, and fairness, in which individual mobility and status permeability are possible. Under this framing, hard work is perceived as leading to success, and failure is framed as a result of personal shortcomings (Eccleston et al. 2010). This belief system is even more common among white Americans and is more likely to be endorsed by those in higher status groups, who have lack of direct experiences that contradict that narrative. Research by Eccleston et al. (2010) after Hurricane Katrina found that white Americans were more likely to blame the victims but that increasing empathetic feelings reduced the negative attitudes toward the victims. However, even efforts to assist the victims through volunteerism serve to perpetuate the treatment of recovery and rebuilding as an individual burden and not as a government obligation. The willingness of so many to step forward and render assistance is used to justify the destruction of the social safety net (Flaherty 2016).

The expectations placed upon disaster survivors, that they show their deservedness, stand in sharp contrast to the immediate rush to promote neoliberal approaches to disaster recovery, through the reduction of safeguards around contracting and other policies intended to privilege free-market approaches to disaster recovery. While impacted families and households are being forced to jump through a variety of bureaucratic hoops to ensure that no fraud is occurring, government contracts and private sector profits are not questioned or scrutinized. For example, within one month after Hurricane Katrina, over 15 contracts had been awarded of over $100 million of which five were over $500 million. In fact, of the $1.5 billion in contracts eventually awarded by the Federal Emergency Management Agency (FEMA), over 80% were no-bid or limited competition (Dyson 2006). This type of contracting also favors larger companies at the expense of local companies, further impacting survivors, as the resources that flow in for recovery are diminished by corporate profits. In fact, despite the frequent arguments that disaster recovery provides economic benefits and fosters job creation, contractors typical do not hire locals and utilize their profits to increase corporate infrastructure, often with headquarters located outside of the impacted areas (Klein 2007b).

The disaster market is a growing industry, with companies entering the marketplace following major disasters such as Hurricane Katrina. This can be seen as an extension of the shifting of the military-industrial complex into a disaster-capitalism complex that profits from disaster functions while competing with the state and non-profits that might otherwise fulfill recovery functions (Klein 2007a). Naomi Klein (2007a) refers to this as disaster capitalism, in which the public sphere is raided following large catastrophes, and disasters are seen as prime market opportunities and as an avenue for exploitation—all existing within a "democracy free zone" where the usual need for consensus does not apply. These issues will be further explored in Chap. 2—Deserving Victims and Post-Disaster Fraud.

A final area for analysis is the impacts of government corruption and ineptitude, both of which further the victimization of those impacted by a disaster through institutional and structural violations of human rights. Examples of government failures, including those directly resulting from corruption, are purchasing inadequate services and products, enabling discriminatory practices, the failure to fully prepare or plan for response, and the failure to protect the public from official corruption (Voigt and Thornton 2015).

1.5 Survivor Agency

The need to prove deservedness severely impacts a person's ability to exercise their agency. This challenge has been recognized in some of the critiques of the term vulnerable, which is described as placing the fault with the individual and not with the system (Jacobs 2018). Fundamentally, vulnerability is not caused directly by race and class; instead, it stems from the impacts of sexism, racism, and classism. These injustices diminish survivor agency, including its expression through collective efficacy in the public sphere, as do many of the programs that are put in place to provide assistance during disaster recovery, including those that are not managed by the government. Some theorists, such as Naomi Klein (2007b), have even argued that disasters are used to purposefully erode the public sphere and further a vision of a neoliberal world in which individuality is the highest value.

One of the means through which individuals are able to exercise their agency is through their connections to their community via social capital, a concept further described later in this chapter. Research has found that strengthening social capital, which increases access to resources and collective efficacy, is key to increasing disaster preparedness and allowing for greater engagement with policy making around recovery (Mathbor 2007). Discussions of the role of social capital in community resilience often look at three levels of social capital: bonding (linkages to similar individuals), bridging (linkages with individuals outside of the immediate group), and linking (connections to institutions). Social systems and social support (including affect, affirmation, and aid) have been shown to reduce the effects of chronic and acute stressors, but it is possible for the demands upon the system to exceed its ability to respond. This is often the case for distressed communities facing an acute shock such as a massive flood. These systems, which might otherwise function to increase resilience, can collapse and be unable to provide needed assistance (Bolin and Bolton 1986). However, framing a community as simply vulnerable detracts from the agency of its members and ignores the pre-existing social supports and social capital. A more just approach to disaster recovery would build upon these existing capacities.

Non-profit organizations and philanthropy are complicit in both the reduction of individual agency and the perpetuation of unjust hierarchies—beginning from the initial premise that the poor or vulnerable need saving, but it is also inherent in their participation within the neoliberal marketplace and the compositions of boards and staff that do not reflect the communities served. Another means through which charitable endeavors serve to perpetuate the status quo is through the creation of a "moral safety valve," legitimizing personal financial generosity, or donations of volunteer hours, as a sufficient response to the temporal expressions of major social problems (Flaherty 2016). In other words, volunteering to help feed a handful of disaster survivors, without engaging more deeply in any efforts at systemic change while the consequences from disasters continue to increase inequalities and disproportionately affect the minority population (Voigt and Thornton 2015).

Individual agency also supports collective action, which can be described as a networked version of human agency. Collective action is critical to the exercise of collective agency, through political mobilization, participation in public processes, or efforts to ensure justice. According to Zakour and Swager (2018), collective action manifests through collective efficacy, creativity and flexibility, problem-solving skills and reflection, community action, and political partnerships. This concept will be further explored in Chap. 3.

1.6 Social Capital and Collective Efficacy

A great deal of literature has explored the benefits of social capital in disaster preparedness, recovery, and resilience. As was previously discussed, social capital is often divided into bonding, bridging and linking capital—drawing a distinction between the types of connections, as well as the groups and individuals with which one is connected. Another way in which social capital is described is in the distinction between structural and cognitive capital, where structural capital stems from formal organizations and networks while cognitive capital involves norms and attitudes that increase social cooperation (Radu 2018). Structural social capital allows communities and families to better access formal planning processes, ensure representation, and advocate for their needs in recovery. Cognitive capital increases social connections and social supports, both of which have been shown to improve resilience and recovery outcomes, as well as to reduce trauma and depression (Radu 2018; Akber and Aldrich 2018; Rung et al. 2017).

Other researchers have discussed the importance of social and cultural resources, such as sense of place, which then correlate with perceptions of neighborhood cohesion and have been found to support recovery and return (Chamlee-Wright and Storr 2009). Social capital and community support can take many forms including information sharing, direct life and property protection, assistance with caregiving, and emotional support (Radu 2018). All of these direct benefits from social capital support the exercise of individual agency in recovery. However, despite the fact that social capital can have positive effects upon resilience, the impacts of a crisis can actually serve to disrupt social cohesion and negate the benefits of social capital when it is most needed (Doherty 2004). The inability to access social networks and support then impedes the formation of a therapeutic community (Gill 2007). This type of disruption can occur for many reasons including competing narratives, questions around responsibility, and the concentration of detrimental impacts upon particular segments of the population (Aronoff and Gunter 1992).

The loss of community cohesion has been described as a secondary trauma, or a secondary disaster, following the initial disaster, where inadequate response, prolonged displacement, and other factors can disrupt the social fabric (Gill 2007). Key components of social capital that can be impacted include social norms, attitudes, trust, and reciprocity (Fleming et al. 2014). Fleming et al. (2014) describe four conditions that lead to a decline in trust and reciprocity, in particular: social displacement within communities, such as when temporary housing is placed in group sites; migration away from the community; information asymmetry between community members; and rivalry or disputes over scarce resources. These conditions are exacerbated by divisive policies that focus on fraud and paint survivors as opportunists.

1.7 Public Policy and Legislation

Disaster policy in the United States involves a complex set of interactions between different policy mechanisms, federal agencies, and across all levels of government. These interactions are taking place within the complex, open system that is a disaster event—one impacting individuals, households, organizations, neighborhoods, institutions, regions, states, and the nation. Furthermore, the complex intergovernmental partnerships that are required for the administration of post-disaster recovery programs, as well as pre-event preparedness and hazard mitigation programs, are strained by budget constraints, increasing frequency of events, and rising losses. The frequency of presidential disaster declarations, with all 50 states and Washington D.C. having received at least one between 2005 and 2014, has resulted in a growing national discussion regarding reducing federal spending and further straining state resources (Pew Charitable Trusts 2018).

One reason for increasing impacts to the built environment is the fact that the United States has not prioritized spending on infrastructure maintenance, due in part to an ideological war on "big government," leading to infrastructure decline and increasing the risk from natural hazards (Klein 2007a). Impacts on individuals and households are also increasing, particularly to renters and low-income homeowners, a population that existing disaster policies and programs are not designed to serve. Current policies, stemming from the Stafford Act, favor homeowners and those with flood insurance, providing much less assistance to renters and those with properties that command a lower value. This has been clearly demonstrated in prior disaster recovery efforts, such as those in New York and New Jersey following Superstorm Sandy, where low-income families and people of color were disproportionately impacted by disaster recovery policies (Maxwell 2018).

In fact, many federal policies serve to perpetuate injustice, often in the name of objectivity. The use of a benefit-cost approach for the allocation of disaster-related assistance, including funding for disaster risk reduction, privileges population centers and areas with a higher value housing stock above those areas that are more impoverished and contain lower value infrastructure or buildings—often reflecting decades of public disinvestment in those same communities. However, the benefit-cost model does not account for the role of values and historic practices; instead, it defines objectivity as freedom from any values and focuses solely on the monetization of risk and impacts through aggregation that cannot factor in distributive inequalities (Shrader-Frechette 2002). A similar concern has been raised within the field of environmental justice, where the risk assessments that are utilized to justify public decision-making quantify benefits in an economic fashion while treating negative impacts as qualitative and, therefore, not able to be included within the equations. These negative impacts are treated as factual and inevitable with little discussion of alternatives or context.

1.8 Implementation Challenges

The size of the United States, the complexity of intergovernmental relationships, the socio-political and geographic differences across regions, the increasing wealth gap, and other complicating factors such as the wide range of capabilities and capacities between and across communities, all serve to make the implementation of disaster recovery challenging at any time. These discrepancies, spread out geographically, render a one-size approach to disaster recovery impossible (Cutter and Emrich 2006). Population migration to areas of higher risk, including development pressures that prioritize growth over safety, further increase the challenges of implementation, particularly for disaster risk reduction. For example, the population of the US population living in a coastal county rose from 28% to 53% between 1980 and 2003 (Cutter and Emrich 2006).

Challenges to implementation are not only a concern where federal programs are concerned, but they are also a challenge when NGOs and community organizations engage in disaster recovery. Despite often having good intentions, the results from prior efforts at supporting community recovery have been varied. Allen (2013) studied the recovery of two different neighborhoods in New Orleans, Holy Cross and Broadmoor, both of which had similar demographic characteristics, had established neighborhood associations, and relied heavily on outside assistance. The study found that although both neighborhoods had been targeted for demolition by the Bring New Orleans Back Commission' redevelopment plan, their success at engagement with NGOs and outside partners was very different. In the case of Holy Cross, the external experts drove an agenda focused on sustainability and with little resident input, resulting in solutions that were provided on a home-by-home basis and ignored economic realities. In Broadmoor, Harvard brought technical resources, but did not lead the effort, and the planning process divided the neighborhood into three parts in order to recognize economic disparities and give all groups a voice (Allen 2013). These findings suggest that representation and participation, both tenets of a justice approach, must be explicitly addressed in order for recovery efforts to be successful, and that justice must be prominently considered in implementation strategies. This will be further discussed in Chap. 5—Implementation Challenges.

1.9 Disaster Risk Reduction and Creation

An economic system that requires never-ending growth and resists regulation has the direct result of generating disasters and disaster impacts through market mechanisms (Klein 2007b). This constant creation of risk takes place alongside the commodification of safety and disaster risk reduction, sold at a premium solely to those who can afford the added costs of a constantly shifting safety.

The prevailing narratives around climate adaptation and hazard mitigation take a technocratic approach, one that privileges data above people, and argues for colorblind risk reduction. Unfortunately, colorblind adaptation practices, utilized in landscapes that were not created through colorblind processes, and in which risk is not distributed in colorblind ways, will only perpetuate existing inequalities and injustices. As Hardy et al. point out (Hardy et al. 2017, p. 62) narratives and hazard mitigation and climate adaptation ignore the "landscapes of race and deep histories of racism that have shaped the socio-ecological formations of coastal regions." Adaptation practices that expect communities of color, and communities of place, to bear the psychological and economic costs of migration, for example, are inherently unjust—asking those who have least benefited from the risks perpetuated by a capitalist and neoliberal system to now bear the brunt of the costs, or "bads," created.

1.10 Disparate Outcomes

Across New Orleans and Southeast Louisiana, marginalized communities were the most impacted by flooding, the least able to participate in recovery, rebuilding, and reconstruction efforts; and in many cases, their recovery was delayed or even permanently disrupted (Zakour et al. 2018). In the ten years following the storm, a full 90% of residents returned to their neighborhoods; however, in the Lower Ninth Ward, a historically African

American community, only 37% of residents were able to return within that same time period. Across the City of New Orleans, there are now 92,000 fewer African American residents than there had been prior to Hurricane Katrina (Maxwell 2018). These disparate outcomes are influenced by a lack of survivor agency, a lack of access to capitals, and a failure to make policy decisions from a PPFPE premise.

Hurricane Katrina is just one example of the disparate recovery outcomes that are currently the norm. Chapter 7—Disparate Outcomes—will provide further examples of these disparate outcomes, which both stem from injustice and perpetuate it.

1.11 Resilience for Whom?

Asking the question, resilience for whom, is imperative if we are to achieve disaster justice. In 2017, the Texas Low Income Housing Information Service, also known as Texas Housers, a non-profit organization established in Austin, put forth seven principles for equitable disaster recovery that provide a starting point for a conversation about Just Recovery. These principles are a direct result of Texas Housers' history of working with low-income and disadvantaged communities across Texas, many of whom have born disproportionate impacts from hurricanes and flooding. The principles draw from the lessons learned from Hurricanes Dolly and Gustav, as well as the ongoing challenges faced in responding to Hurricane Harvey (Rosales 2007).

The principles include ensuring that government assistance is provided in a timely fashion, easy to understand, and accessible; temporary housing is made available to everyone in a way that allows for a reconnection to family; those who have been displaced have access to all needed resources, and disaster rebuilding provides local employment at fair wages; survivors have a say in policy decisions and all programs are transparent and meeting civil rights laws; and all homeowners and renters are able to rebuild in ways that fit their needs, in neighborhoods of their choice, with equal investments in infrastructure, and that are free from environmental hazards.

These seven principles provide a roadmap to ensuring a more just approach to resilience and Just Recovery, addressing many of the challenges and barriers to individual agency identified within this book and providing a direct linkage to the incorporation of a justice framework into recovery processes. In summary, free and informed consent is not possible when there is a lack of information or when information is presented at different points in time leaving community members unable to make an informed decision and choice is a critical component of well-being; ensuring that basic needs are met, and that housing improves access to social capital and community networks, is critical to maximizing individual agency; unequal access to resources, and discriminatory practices in resource allocation and infrastructure investment would have to be explicitly justified utilizing a PPFPE approach; and full participation, in ways that increase agency and do not exacerbate trauma, including access to decision-makers and the resources for informed decision-making, is a key premise of Just Recovery.

1.12 Defining Just Recovery

This text proposes an initial set of practices and principles that characterize the notion of a Just Recovery. These principles are only a starting point and will require further elaboration and operationalization in order to be fully translated into actionable guidelines for practice. Accomplishing any one of these principles fully would require extensive changes to existing policies, as well as to the practices utilized in their implementation. This text is not intended to provide a full roadmap but is intended to further scholarly and practical discussions regarding the ways in which justice can be achieved in disaster recovery.

Principle #1: Just Recovery requires that all community members (regardless of their socio-economic status, race, gender, sexual identification, land tenure, etc.) be provided with the ability to exercise their agency fully through free and informed choice in support of their personal well-being. Full and informed choice is not possible if there is any coercion, exclusion from public policies, or other barriers to full participation. Furthermore, agency cannot be fully exercised if a full and complete array of options is not understood and made available within a timely fashion and in means that are accessible.

Principle #2: Just Recovery begins from the PPFPE, which clearly establishes that any different or unequal treatment must be justified by the discriminator; only equality is inherently defensible. Any expectation that disaster victims or communities prove their deservedness in the face of impartial bureaucratic processes puts the onus on the victims to justify the need for equal treatment and fails the PPFPE test.

Principle #3: Just Recovery requires the full harnessing of the communities' transformative and adaptive capacity, honoring their definitions of resilience, in order to reduce risks for the future. Holistic disaster risk reduction is not possible without acknowledging existing patterns of unequal distribution of risk. It is not sufficient to simply mitigate against some current risks in rebuilding; instead underlying structures and patterns must be questioned. Colorblind and a-historic recovery, which does not consider context, is not Just.

Principle #4: Just Recovery is not possible without equal access to resources and programs, including full participation in decision-making processes that govern resource allocation, future development, and other related functions.

Notes

1. This internal migration could reasonably be considered to have been a forced migration, driven by recovery policies that substantially limited the abilities of some residents, such as those in public housing, to return to their original homes or even to their community.
2. It should be noted that no disaster is fully natural, as socio-cultural processes create and exacerbate vulnerability leading to disaster impacts that would not occur solely as a result of the hazard itself. This book will avoid the use of the term natural disaster, in recognition of this fact.
3. It is important to consider intersectionality in any discussion of identity, as individuals have a range of identities and experiences that combine in different ways.
4. Although forced migration is a strong term, and one that brings to mind human rights violations, it can be argued that policies and programs which leave no alternatives to leaving by limiting access to adaptation, do in fact force migration.
5. The term natural disasters is used here, as it is the terminology utilized for the analysis and denotes the limitations in the types of hazards considered for the statistic.
6. The term Just Recovery is capitalized in order to distinguish a justice-based framing of recovery from one which merely references justice among various parameters.
7. Uneven distribution or unjust/unfair distribution.
8. Zakour and Swager (2018, pp. 63–64):

    1. "The vulnerability of social systems is the reduced capacity to adapt to environmental circumstances;
    2. Vulnerability is not evenly distributed among people or communities;

3. The concept of disaster vulnerability is multidimensional;
4. The availability and equitable distribution of resources in a community decreases disaster vulnerability and facilitates resilience;
5. Vulnerability is largely the result of environmental capabilities and liabilities;
6. Social and demographic attributes of people are associated with, but do not cause, disaster vulnerability;
7. Unsafe conditions in which people live and work are the most proximate and immediate societal causes of disaster;
8. Root causes, the socio-cultural characteristics of a community or society, historically and in the present, are the ultimate causes of disasters;
9. Disasters occur because of a chain of causality in which root causes interact with structural pressures to produce unsafe conditions. Hazards then interact with unsafe conditions to trigger a disaster;
10. Culture, ideology, and shared meaning are of central importance to the progression of disaster vulnerability;
11. Environmental capabilities and liabilities, and disaster susceptibility, are related in complex ways to produce the level of community vulnerability; and,
12. The environments of communities are growing in complexity and are increasingly global in scale."

References

Adriano, L. (2018, February 12). Experts: As Floods Worsen, So Does the Risk of Toxic Spills. *Insurance Business America*.

Akber, M. S., & Aldrich, D. P. (2018). Social Capital's Role in Recovery: Evidence from Communities Affected by the 2010 Pakistan Floods. *Disasters, 42*(3), 475–497.

Allen, B. C. (2007). Environmental Justice and Expert Knowledge in the Wake of a Disaster. *Social Studies of Science, 37*(1), 103–110.

Allen, B. C. (2013). Justice as Measure of Nongovernment Organization Succeed in Post-Disaster Community Assistance. *Science, Technology, and Human Values, 38*(2), 224–249.

Aronoff, M., & Gunter, V. (1992). It's Had to Keep a Good Town Down: Local Recovery Efforts in the Aftermath of Toxic Contamination. *Industrial Crisis Quarterly, 6*, 83–97.

Bolin, R., & Bolton, P. (1986). *Race, Religion, and Ethnicity in Disaster Recovery*. Program on Environment and Behavioral Science, University of Colorado.

Chamlee-Wright, E., & Storr, V. H. (2009). "There's No Place Like New Orleans": Sense of Place and Community Recovery in the Ninth Ward After Hurricane Katrina. *Journal of Urban Affairs, 31*(5), 615–634.

Cutter, S. L., & Emrich, C. T. (2006, March). Moral Hazard, Social Catastrophe: The Changing Face of Vulnerability Along the Hurricane Coasts. *ANNALS, AAPSS, 604*, 102–112.

Doherty, G. W. (2004, September). Crisis in Rural America: Critical Incidents, Trauma and Disaster. *Traumatology, 10*(3), 145–164.

Dyson, M. E. (2006). *Come Hell or High Water: Hurricane Katrina and the Color of Disaster*. New York: Basic Civitas Books.

Eccleston, C. P., Kaiser, C. R., & Kraynak, L. R. (2010). Shifts in Justice Beliefs Induced by Hurricane Katrina: The Impact of Claims of Racism. *Group Processes & Intergroup Relations, 13*(5), 571–584.

Edwards, G. A. S., Reid, L., & Hunter, C. (2016). Environmental Justice, Capabilities, and the Theorization of Well-Being. *Progress in Human Geography, 40*(6), 754–769.

Federal Emergency Management Agency. (2018, July 12). *2017 Hurricane Season FEMA After-Action Report*.

Fink, S. (2018). Nearly a Year After Hurricane Maria, Puerto Rico Revises Death Toll to 2,975. *The New York Times*. Retrieved from https://www.nytimes.com/2018/08/28/us/puerto-rico-hurricane-maria-deaths.html.

Flaherty, J. (2016). *No More Heroes: Grassroots Challenges to the Savior Mentality*. Chico, CA: AK Press.

Fleming, D. A., Chong, A., & Bejarano, H. D. (2014). Trust and Reciprocity in the Aftermath of Natural Disasters. *The Journal of Development Studies, 50*(1), 1482–1493.

Freudenburg, W. R., et al. (2008). Organizing Hazards, Engineering Disasters? Improving the Recognition of Political-Economic Factors in the Creation of Disasters. *Social Forces, 87*(2), 1015–1038.

Gill, D. A. (2007). Secondary Trauma or Secondary Disaster? Insights from Hurricane Katrina. *Sociological Spectrum, 27*, 613–632.

Goodell, J. (2018, February 25). Welcome to the Age of Climate Mitigation. *Rolling Stone*.

Hardy, R. D., Milligan, R. A., & Heynen, N. (2017). Racial Coastal Formation: The Environmental Justice of Colorblind Adaption Planning for Sea-Level Rise. *Geoforum, 87*, 62–72.

Hernandez, A. R., & McGinley, L. (2018, May 29). Harvard Study Estimates Thousands Died in Puerto Rico Because of Hurricane Maria. *The Washington Post*.

Jacobs, F. (2018). Black Feminism and Radical Planning: New Direction for Disaster Planning Research. *Planning Theory, 00*(0), 1–16.

Kelly, C., Costa, K., & Edelman, S. (2017, October 3). *Safe, Strong, and Just Rebuilding After Hurricane Harvey, Irma, and Maria Recommendations*. Center for American Progress.

Klein, N. (2007a). Disaster Capitalism, The New Economy of Catastrophe. *Harper's Magazine, 315*(1889), 47–58.

Klein, N. (2007b). *The Shock Doctrine*. New York, NY: Picador.
Mark, J. (2018, April 23). The Case for Climate Reparations. *Sierra Club Magazine*.
Mathbor, G. M. (2007). Enhancement of Community Preparedness for Natural Disasters: The Role of Social Work in Building Social Capital for Sustainable Disaster Relief and Management. *International Social Work, 50*(3), 357–369.
Maxwell, C. (2018, April 5). *America's Sordid Legacy on Race and Disaster Recovery*. Center for American Progress.
Norton, R., MacClune, K., Venkateswaran, K., & Szönyi, M. (2018). *Houston and Hurricane Harvey: A Call to Action*. Zurich, Switzerland: Zurich Insurance Company Ltd.
Ottinger, G. (2012). Changing Knowledge, Local Knowledge, and Knowledge Gaps: STS Insights into Procedural Justice. *Science, Technology, & Human Values, 38*(2), 250–270.
Palm Beach Post. (2018, June 4). Editorial: Maria's Deadly Impact on Puerto Rico Deserves More Attention.
Pelling, M. (2003). *The Vulnerability of Cities: Natural Disasters and Social Resilience*. Sterling, VA: Earthscan.
Pew Charitable Trusts. (2018, June). What We Don't Know About State Spending on Natural Disasters Could Cost Us: Data Limitations, Their Implications for Policymaking, and Strategies for Improvement.
Radu, B. (2018). Influence of Social Capital on Community Resilience in the Case of Emergency Situations in Romania. *Transylvanian Review of Administrative Sciences*, (54), 73–89. https://doi.org/10.24193/tras.54E.5.
Rivera, J. D., & Miller, D. M. S. (2007). Continually Neglected: Situating Natural Disaster in the African American Experience. *Journal of Black Studies, 37*(4), 502–522.
Rosales, C. (2007, September 14). We Want Disaster Recovery to Be Fair and Just. Here's a Good Place to Start.
Rung, A. L., Gaston, S., Robinson, W. T., Trapido, E. J., & Peters, E. S. (2017). Untangling the Disaster-Depression Knot: The Role of Social Ties after Deepwater Horizon. *Social Science & Medicine, 177*, 19–26.
Shrader-Frechette, K. (2002). *Environmental Justice: Creating Equality, Reclaiming Democracy*. Oxford: University Press Oxford.
Shrader-Frechette, K. (2012, June). Nuclear Catastrophe, Disaster-Related Environmental Injustice, and Fukushima: Prima Facie Evidence for a Japanese 'Katrina'. *Environmental Justice, 5*(3), 133–139.
Swim, J. K., & Bloodhart, B. (2018). The Intergroup Foundations of Climate Change Justice. *Group Processes and Intergroup Relations, 21*(3), 472–496.
Terruso, J. (2017, October 18). How Hurricane Sandy Became Steroids for Jersey Shore Development. *Philly News*.

Tierney, K. (2010). Growth Machine Politics and the Social Production of Risk. *Contemporary Sociology, 39*(6), 660–663.

Voigt, L., & Thornton, W. E. (2015). Disaster-Related Human Rights Violations and Corruption: A 10-Year Review of Post-Hurricane Katrina New Orleans. *American Behavioral Scientists, 59*(10), 1292–1313.

Wisner, B., Blaikie, P., Cannon, T., & Davis, I. (2005). *At Risk: Natural Hazards, People's Vulnerability, and Disasters* (2nd ed.). New York: Routledge.

Zakour, M. J., & Swager, C. M. (2018). Vulnerability-Plus Theory: The Integration of Community Disaster Vulnerability and Resiliency Theories. In M. J. Zakour, N. B. Mock, & P. Kadetz (Eds.), *Creating Katrina, Rebuilding Resilience: Lessons from New Orleans on Vulnerability and Resiliency* (pp. 45–73). Oxford: Butterworth-Heinemann.

Zakour, M. J., Mock, N. B., & Kadetz, P. (2018). Editor's Introduction: The Voices of the Barefoot Scholars. In M. J. Zakour, N. B. Mock, & P. Kadetz (Eds.), *Creating Katrina, Rebuilding Resilience: Lessons from New Orleans on Vulnerability and Resiliency* (pp. 3–23). Oxford: Butterworth-Heinemann.

CHAPTER 2

Deserving Victims and Post-Disaster Fraud

Abstract This chapter focuses upon the construct of the deserving victim, as seen in the context of natural and man-made disasters, as well as in other social welfare contexts. This concept is applied at the community scale as well, with the onus being placed upon the community to prove its deservedness. The chapter seeks to integrate findings from fields such as psychology and sociology relative to individual and cultural discomfort with the notion of allowing individuals to "cheat," with the literature around disaster recovery. In other words, the chapter interrogates the underpinnings of the extensive focus on fraud that has characterized post-disaster assistance and aid dating back to the creation of those mechanisms and contrasts it with the assumption of corporate deservedness—as shown by the typical failure to question high corporate profit models and the unwillingness to hold corporations to account for anything less than the most egregious violations.

Keywords Privatization • Neoliberalism • Post-disaster fraud
• Corrosive community • Altruistic community • Disaster capitalism

This chapter focuses upon the construct of the deserving victim, as seen in the context of disasters, as well as in other social welfare contexts. It seeks to integrate findings from fields such as psychology and sociology relative to individual and cultural discomfort with the notion of allowing individuals to "cheat," with the literature around disaster recovery and social capital.

In other words, the chapter will interrogate the underpinnings of the extensive focus on fraud that has characterized post-disaster assistance and aid dating back to the creation of those mechanisms and contrast it with the assumption of corporate deservedness—as shown by the ongoing failure to question high corporate profit models and the unwillingness to hold corporations to account for anything less than the most egregious violations. This failure will be illustrated through various examples from Hurricanes Katrina, Sandy, and Maria.

The majority of the post-disaster assistance to individuals and communities within the United States is provided through programs that are managed by the Federal Emergency Management Agency (FEMA) and the Department of Housing and Urban Development (HUD). Additional programs from several other federal agencies can also be brought to bear, as can state-level programs and other sources of assistance.[1] Each of these programs has extensive rules and regulations in place, many geared toward reducing the duplication of benefits or fraud but amounting to a set of procedures that can place an undue burden of proof on recipients (Laska et al. 2018; Hammer 2015).

Despite a consistent cultural and agency narrative that highlights the misuse of funds, the vast majority of financial losses due to "fraud" following a disaster can actually be attributed to programmatic and contractor errors, not to intentional theft, as can be shown by a closer exploration of the analytical techniques utilized by agency reports (OIG 2015). However, the institutional processes and narratives create a fear of fraud that leads to extensive expenditures and hinders the ability to deliver aid to those who most need it while discouraging potential recipients from seeking aid, contributing to both the information asymmetry and perceptions of differential levels of aid that hinder social capital (Reid 2013). This has been documented through case studies of case management and disaster recovery programs following both Hurricanes Katrina and Superstorm Sandy (Jerolleman et al. forthcoming).

Additionally, the high profit margins and layers of contracting that have become the norm take far more money away from survivors than does fraud. At the same time, the focus on fraud supports the creation of a corrosive community, a term historically used to describe a pattern of chronic negative social impacts to individuals and communities that impede recovery in the aftermath of a technological disaster or toxic contamination (Cuthbertson and Nigg 1987; Freudenburg 1997; Picou 2009; Brunsma et al. 2010). The corrosive community impedes the formation of a therapeutic community and the utilization of social bonds for recovery.

Though most commonly used for technological disasters, Peacock and Ragsdale (2000) also noted that unconstrained competition between organizations, agencies, and local community groups following disasters can seriously impede community recovery and lead to prolonged negative social impacts. Prominent examples where a corrosive community has been demonstrated include the Exxon Valdez Oil Spill (Picou 2009), the Deepwater Horizon Spill (Mayer et al. 2015), and the Buffalo Creek Disaster (Scott et al. 2012). Corrosive community has also been studied in relation to Hurricane Katrina, a notable disaster in that it can be categorized as technological despite having involved a natural hazard (Brunsma et al. 2010).

2.1 Deserving Victims

Media and other public portrayals of victims as opportunistic looters or violent criminals, as were seen following Hurricane Katrina and are routinely seen when communities of color are impacted, are both a result of existing biases[2] regarding people of color, based upon social value judgments, and an amplifying factor, supporting elite panic and impacting allocations of public safety resources and public empathy (Tierney et al. 2006). These portrayals of increased crime and violence also serve to justify a military response to disasters, focused on property protection above justice, and ignore the historic context behind the creation of vulnerability. These narratives place the victim at fault for their own risk due to either moral failings or a perceived unwillingness to comply with government directives such as evacuation order.

Families and households impacted by disasters are expected to provide extensive documents of ownership[3] in order to receive financial assistance; this is particularly challenging in communities where there is a great deal of informal housing, as is the case in Puerto Rico (where around half of the homes were built without permits and lack titles), or where a long history of redlining led to a prevalence of bond for deed credit sales and other informal mechanisms of transferring ownership (Adams 2013; Florido 2018). In fact, of the 1.2 million applicants in Puerto Rico, over half were found ineligible for assistance due to an inability to verify ownership.[4] Foreign Exchange Management Act (FEMA) representatives have responded to advocates by stating that documentation requirements have been eased, but many applicants continue to be denied and there is inconsistency in terms of what is considered acceptable documentation (Florido 2018).

The portrayals of disaster victims, particularly those in minority groups such as Latino Workers[5] (a population which increased three-fold following Hurricane Katrina), by the media and in public discourse, also served to increase the vulnerability of these populations to injustices such as wage theft and other dangers inherent in the recovery process, such as health impacts. The dehumanization of workers, portrayed as workhorses in highly racialized terms, allowed corporations to place these individuals at high levels of personal danger while maximizing their profits. Many Latino workers were left to do the most dangerous and hazardous jobs while facing threats when they demanded the payment of wages and being targeted by police officers (Trujillo-Pagan 2012). One estimate by the Southern Poverty Law Center showed 80% of Latino workers had been ripped off by employers while in New Orleans. These crimes, and others, were either unreported or simply not a priority for law enforcement (Voigt and Thornton 2015). The general public discourse regarding their presence in the city was one of discomfort, heightened by their portrayal as criminals, despite the fact that the workers themselves were targeted for robbery due to their inability to open bank accounts and the subsequent reliance on cash.

Disaster assistance mechanisms, including eligibility criteria, are often portrayed as being colorblind, much like adaptation policies and other technocratic policies, but there are clear patterns regarding where and when assistance is, or is not, approved. Following Hurricane Harvey, for example, 34% of impacted white residents were approved for assistance, compared to only 13% of black residents (Dupuy 2017). This pattern of rejection of applications for assistance in the most distressed communities is a reflection of both the policies themselves and their implementation. It serves to perpetuate unjust outcomes, as those who most require assistance are least able to access it. Continuing with the example of Harvey, 65% of Hispanics and 46% of black residents experienced income disruption, as opposed to only 31% of white residents, while having greater difficulty accessing assistance (Dupuy 2017). This is further discussed in the following chapters.

2.2 Duplication of Benefits and Appeals

The various federal programs that are involved in disaster recovery all have procedures in place to minimize the occurrence of fraud, often at high expense to the federal government as layers of oversight add layers of contracting for reviews and other mechanisms. It can even be argued that in some cases the federal government is spending more money trying to prevent fraud through disaster assistance than might otherwise have been lost through the course of some small amount of error or even minimal fraud, although a full analysis is not possible with the data available at this time (Jerolleman et al. forthcoming). Investigations following major disasters rarely document extensive intentional fraud committed by individuals, although some cases of fraud are always discovered. There is, however, often evidence of errors in case management and the administration of programs, stemming from challenges in data sharing and the difficulties inherent in integrating different agency programs (across differing timelines and with different regulations), often resulting in the duplication of benefits. It is often the case that evidence of potential overpayment is found in a certain percentage of cases, and is lumped into the category of fraud, regardless of the reasons for the overpayment. Digging into the data further, and distinguishing between intentional and unintentional duplications of benefits, shows a much smaller amount of potential fraud. For example, in 2015, the Office of the Inspector General found possible duplication in under 0.01% of the 182,000 applications reviewed for potentially false claims of "no insurance" (OIG 2015).

Despite the complexities and errors inherent in the programs, the oversight mechanisms regularly place the undue burden of proof on recipients and can lead to high attrition rates (Laska et al. 2018). This concern was raised following Superstorm Sandy, when FEMA faced scrutiny for its handling of flood claims and verification, often underestimating damages and underpaying claims (in an effort to show reduced costs to the program) while homeowners received conflicting information regarding procedures and appeals (Frontline 2016). The appeals process, itself, is quite lengthy and requires access to resources and information that many individuals lack. Placing the burden of appeals directly upon the impacted households further decreases their ability to secure adequate assistance for recovery while privileging those families that have the income and resources to obtain legal assistance. All the while, corporations tasked with project management and case management continue to charge for their services (even if they are the source of the errors) with unquestioned profit margins. Another example can be found following Hurricane Katrina, where the profusion of red tape proved traumatic for many survivors, in some cases entirely derailing efforts at recovery while failing to prevent errors and then requiring applicants to return funds several years later—placing the burden of the errors directly on those who could least afford to pay it. Some of these errors were due not just to the accidental duplication of benefits but also to the long timelines associated with receiving final awards, resulting in confusion regarding what funds could be retained and what should be returned (Jerolleman et al. forthcoming). FEMA has been working to overhaul the recoupment process in order to better insure the recoupment of improper assistance, but there has not been a concurrent effort to reduce the incidence of duplication of benefits. Instead, the focus has been on debt collection efforts.

2.3 Deserving Victims, Fraud, and the Corrosive Community

The study of cheating within the field of psychology, including the concepts of altruistic punishment, and the role that in-group vs. out-group perceptions play, is useful for an understanding of the socio-political approach to fraud prevention—and its connection to the concept of a deserving victim. The concept of altruistic punishment indicates that individuals feel motivated to punish cheating, even when there is a personal cost involved. In fact, cooperation breaks down in cases where there is no mechanism to punish cheating (Fehr and Gächter 2002). Seen through this lens, the willingness to spend more funds on fraud prevention than might be lost if fraud were allowed to occur seems an offshoot of the inherent human desire to ensure that cheating is not permitted, even when the costs of doing so are high. There is also a concern described in some of the literature around corruption that if fraud is permitted then more fraud will be encouraged. By this argument, the limited amount of fraud is in fact partially due to the knowledge that fraud prevention efforts are included in the disaster policy.

Whether this is in fact the case, or not, is beyond the scope of this text. However, the focus on preventing cheating, and the narrative regarding the deservedness of assistance, can be found as far back as the origins of the federal programs, such as the initial decision to utilize Small Business Administration (SBA) loans following hurricane Betsy (Horowitz 2014). The immediate aftermath of Hurricane Betsy resulted in a call for grants by some legislators in order to assist the Lower Ninth Ward due to the extent of the impacts. The Lower Ninth Ward, even then a predominantly African American neighborhood, was not seen as worthy of assistance, with the explicit suggestion that giving money to the poor would lead to dependence. As a result, direct aid was determined to be an unacceptable option and instead disasters loans were offered at a 3% interest rate. As remains the case to this day, many residents were unable to quality, with 8000 of the 33,000 applications rejected, and all required to go through the processes of fingerprinting and producing deeds. Those that did qualify found themselves going from having no mortgage to owing the federal government, described by some as a return to indentured servitude, a tremendous injustice for an African American community that had struggled to secure its place in the City. The permissible uses of the funds were extremely limited, again in an effort to prevent fraud, resulting in an inability to utilize the funds for recovery where they were most needed. For example, one homeowner was penalized for using a portion of the funds to purchase a vehicle that was needed in order to commute to a job and then repay the loan (Horowitz 2014).

The second concept that can be brought in from the literature on psychology is that of the in-group vs. the out-group. This states that individuals are less willing to trust persons who they perceive as being different from themselves, something clearly seen in the preceding example where congressmen did not feel that the residents of the Lower Ninth Ward could be trusted with disaster assistance. This is also seen in the often inaccurate perceptions of high levels of cheating in welfare, despite there being plentiful data showing that there is not in fact that much prevalence of cheating. As a side note, it is interesting to observe that the lack of psychological connection to those at risk, or vulnerable, from climate change has been shown to facilitate the minimization of any sense of responsibility or social injustice (Swim and Bloodhart 2018). These concepts therefore apply beyond simply this immediate context of US disaster policy.

The focus on disaster fraud, described above, including the widespread narrative of cheating, erodes group support and diminishes individual agency, as the distrust that is perpetuated can lead to the creation of a corrosive community. Ibañez et al. (2003) studied how disaster humanitarian differed between groups of Latino communities, following Hurricane Andrew, and found that resource allocation played a critical role in influencing social conflict. In other words, the perceptions of unequal resource distribution damaged community dynamics and limited the exercise of community efficacy.

The connection between resource allocation, agency, and conflict is explored by Laska et al. (2018) in their recent examination of the impacts of structural violence and social vulnerability on disaster recovery and individual agency. Following Hurricane Katrina and Superstorm Sandy, as is the case following the majority of disasters, institutional processes converged to create two classes of recovery aid recipients, the "deserved" and the "undeserved." Fraud prevention efforts, focused on the "undeserved," severely hindered the ability to deliver aid to the victims in greatest need. One example of this dynamic is in the delays in assistance following these two events. In both cases, delays impacted lower-income survivors, female-headed households, and the otherwise politically disenfranchised. The complexity of the processes prevented or delayed aid from being distributed to those most in need and also discouraged eligible individuals from applying for federal recovery aid from fear of committing fraud (Reid 2013) (Box 2.1).

Box 2.1 Deserving Victims and Federal Assistance: The Case of Puerto Rico
Case Author: Leila Darwish

Assistance from FEMA, particularly through the Individuals and Households Program (IHP), is critical for Americans needing financial assistance to rebuild and recover from disasters. However, getting federal disaster assistance to rebuild homes damaged by Hurricane Maria has not been forthcoming for hundreds of thousands of residents in Puerto Rico, whose applications have faced high levels of rejection.

As of May 1, FEMA had received 1.1 million registrations to the IHP program for disaster assistance. The number of approved registrations for FEMA grants totals 452,290, while the number of registrations deemed ineligible was 335,748. For those who appealed FEMA's decision, the agency has either denied or not answered 79% of the appeals. A recent Oxfam report found that the ratio of grants to total registrations appeared lower than for other hurricanes in the United States (Oxfam Research Report 2018). One of the major reasons FEMA has cited for the high number of declines for Puerto Ricans seeking individual assistance post Hurricane Maria has been that applicants have been unable to prove home ownership. In order to be eligible for FEMA aid under the individual assistance program, applicants need proper documents, and if the agency is unable to prove the applicant's identity, occupancy, or home ownership status, their application will be denied.

(continued)

Box 2.1 (continued)

Puerto Rico's government estimates that more than half of the houses on the island were built informally, without permits or proper legal documentation. Illegal construction, poor housing development practices, unique property laws that differ from other US states, and the inheritance of property without proper documentation have all created a perfect storm, resulting in an unprecedented number of residents lacking the documentation necessary to appease FEMA's long list of eligibility requirements. There are tens of thousands of Puerto Ricans living in housing on lands that began as squatter settlements and grew to become more formal communities over the years. The majority of these developments lack permits, and their lower-income residents lack the appropriate legal documentation to get their applications for federal disaster assistance approved. There are also many applicants who either inherited land from a family member or were allowed to build housing on the property of a family member or friend, again without the proper title or deeds to prove ownership.

Though designed to reduce the occurrence of individual fraud, FEMA's eligibility and documentation requirements have in this case failed to adequately recognize the complexity of the housing situation in Puerto Rico and the unique challenges faced by many residents of Puerto Rico in providing the proper documentation. While FEMA has taken some steps to ease its documentation requirements (such as accepting sworn affidavits from people who lack a housing deed), applicants continue to be denied and hundreds of thousands of Puerto Ricans still remain in limbo with their applications for individual disaster assistance as they try to reconstruct their homes and rebuild their lives more than a year after Hurricane Maria.

2.4 Deserving Corporations

Several types and scales of corporations and private companies play a role in disaster recovery. These include both the larger corporations that work as contractors, directly to the federal government, and the many layers of subcontracts down to small firms. An additional player in the private sector is the small contractors that work directly with impacted organizations and families. Although there is extensive scrutiny where individual fraud is concerned, there is much less attention paid to the mismanagement of funds by contractors or to the processes by which contracts are granted following disasters. At the individual level, many households fall victim to fraudulent contractors, particularly the elderly and others who do not fully understand the disaster assistance system and fall prey to unethical practices. In fact, one reason why some households are unable to complete repairs despite receiving insurance proceeds and other assistance is that they fall victim to this type of fraud. At times, this also leads to the accidental duplication of benefits when multiple sources of funds are utilized for the same elements of rebuilding due to the loss of some of those funds to fraud. Some contractor fraud can be documented through an analysis of civil lawsuits, but much of it goes undocumented and even unreported. This chapter will focus primarily on the role of corporations, and corporate profiteering, in disaster response, but it is important to note that the discussion regarding fraud does not tend to fully capture these kinds of incidents.

Disaster response, and other activities involved in disaster risk reduction and preparedness, takes place in the context of broader trends regarding the role of government, the privatization of social services, and the role of the private sector in governance. The reconfiguration of governance models over time, dating back to the privatization and New Public Management movements of the 1980s, has led to a shift where the government is not a purchaser of services and often not a direct provider. This shift requires a different skill set for public administrators, in which the state must have the capacity to oversee contractors, ensure accountability, and carry out oversight (Fine et al. 2016). Like many other federal agencies, FEMA manages contracts and diverse networks through both formal and informal mechanisms. In particular, FEMA has outsourced a good bit of its logistics efforts in recovery, based upon the notion that the private sector is better equipped to quickly provide resources, and despite limitations in the ability to manage grants and contracts (Gotham 2010). However, as the 2017 hurricane season illustrated with examples such as the utility repair contracts following Hurricane Maria, this is not necessarily the case. In many cases, the private sector is also ill-equipped to respond to large-scale catastrophes and unable to provide the manpower and resources needed in a timely and cost-reasonable fashion. In other cases, relief efforts appear directly designed to enrich well-connected corporations (Calhoun 2006).

The assumption that corporations will be utilized across all phases of emergency management, including an assumption that profit is a permissible component of privatization, has been perpetuated by federal frameworks and grant mechanisms that assume that contractors will be utilized and in some cases facilitates the disbursement of funds for contracts over the reimbursement for direct services (Jerolleman 2013). A common governmental response to catastrophic disasters has been to immediately reduce any perceived barriers to a corporate response. Neoliberal policies are immediately brought to bear in order to allow for the use of contracts and sub-contracts within disaster recovery, prioritizing private business interests over labor protections and affirmative action policies. Following Hurricane Katrina, for example, the Department of Homeland Security (DHS) issued a 45-day moratorium on employee sanctions for I-9 violations, at the same time that President Bush suspended the Davis-Bacon Act for two months. The stated intent was to save taxpayers' money, but there were no direct requirements that savings be passed on to the government instead of being simply absorbed into corporate profits (Voigt and Thornton 2015). This increased worker vulnerability and facilitated the exploitation of the undocumented Latino labor force. The Occupational Safety and Health Administration (OSHA) and the Institution of Occupational Safety and Health (IOSH) did not elect to enforce worker safety regulations, blaming the scale of the event for their inability to do so (Trujillo-Pagan 2012).

In spite of the predominance of contracting, oversight remains a significant challenge for states, particularly where large contracts such as those that are utilized for disaster recovery are concerned. Research into oversight practices in New Jersey, following Superstorm Sandy, found that the state neglected both performance management and accountability, in part due to an inability to effectively monitor the contractor and in spite of contract provisions that required weekly progress reports and monthly status reports. In fact, eight months into one contract, over $51 million of a $67 million, three-year contract had been expended and no progress reports had been submitted (Fine et al. 2016).

A more recent example is the $300 million contract to rebuild Puerto Rico's power grid that was awarded through a no-bid contract to Whitefish Energy Holdings LLC, a small firm based in Montana with only two employees. This contract was awarded in spite of the existence of a strong mutual aid network between utilities, with no clear explanation from the Puerto Rico Electric Power Authority (PREPA) as to the reasoning behind the award (Flavelle et al. 2017). In this particular instance, the contract was investigated by the Office of the Inspector General and eventually rescinded, but not before having led to significant delays in the restoration of power and forcing the mayor of San Juan to spend time advocating against the contract instead of focusing on other recovery needs. The extended power outage contributed to the death toll, showing that corporate greed and corruption can directly lead to deaths (Sorensen and Horrow 2018). More often than not, these types of contracts have gone unquestioned, but in this case, a 2015 audit of FEMA grants, showing duplication of payments, improper costs, and failure to follow guidelines among grant recipients, appears to have increased the pressure on the agency (Flavelle et al. 2017). It remains to be seen whether this increased pressure lasts and whether it is born to bear in areas such as Texas.

The assumption that the private sector can be held accountable in ways that the government cannot does not hold true in cases of market failure, where the government is not well informed and changing producers is very difficult. Some public administration theorists have proposed relational contracting as an alternative model, in which a mutually beneficial principal agent relationship is created. However, this model is resource intensive, requires a lot of contract management skills, and is often impossible under current conditions and regulations (Fine et al. 2016). Regardless of the contracting model utilized, systems that prioritize profit and rely on the market have no incentive to prioritize justice and the care of vulnerable populations (Zakour and Grogg 2018) (Box 2.2).

Box 2.2 Deserving Corporations: The Case of Puerto Rico
Case Author: Leila Darwish

Big disasters are big money when it comes to contractors and private companies bidding on multi-million dollar FEMA contracts. Following Hurricane Maria, private companies were given lucrative contracts to assist with key components of hurricane response and recovery. However, some of these contractors lacked previous disaster experience and sufficient staff, infrastructure, and resources to fulfill their contacts. The fallout of this corporate greed and ineptitude was critical delays that meant essential goods and services went undelivered to millions of Puerto Rican residents in need. Three of the most notable and controversial examples of this included:
- **Tribute Contracting**, a one-person Atlanta-based company that lacked disaster relief experience, was awarded a nearly $156 million contract by FEMA to provide 30 million ready-to-eat meals to

(continued)

Box 2.2 (continued)

Puerto Rico by October 23, 2017. Tribute only delivered 50,000 of those meals, and the contract was terminated 20 days later. Further investigation revealed that the company had a history of problems with government contracts, including cancelations, and had also been barred from government work until 2019.
- **Bronze Star LLC**, a newly created Florida company with an unproven record of disaster relief and minimal staff and resources, was awarded more than $30 million in contracts from FEMA to supply 500,000 emergency tarps and 60,000 rolls of plastic sheeting for home repairs. The company had been formed by two brothers in August 2017 and had never before won a government contract or delivered tarps or plastic sheeting. FEMA eventually terminated the contracts when after nearly four weeks, Bronze Star had failed to deliver any of the much-needed tarps to residents in Puerto Rico.
- **Whitefish Energy**, a Montana-based company with only two staff, was awarded a controversial $300 million contract by the PREPA to restore electricity to Puerto Rico's 3.4 million residents after Hurricane Maria. At the time, Whitefish Energy retained no linemen on staff and had never been involved in a disaster scenario. Though standard procedure for near-term disaster response is for utilities to enter into mutual aid agreements with their counterparts in other states, the PREPA chose instead to hire a for-profit company to accomplish the monumental task of rebuilding Puerto Rico's electrical grid. However, further investigation revealed that Whitefish was overcharging PREPA for labor and that their contract included clauses making it hard to end the contract or for the government to conduct audits. The contract was canceled amidst intensifying public outcry and political investigation. Under the contract, Whitefish was charging $330/hour for a site supervisor and $227.88/hour for a "journeyman lineman." Subcontractors, who make up most of Whitefish's workforce, charged $462/hour for a supervisor and $319.04/hr for a lineman. It also came to light that US Interior Secretary Ryan Zinke had personal and familial connections to the company, which was located in his hometown.

2.5 Government Corruption

Another reason often cited for the creation of complex anti-fraud mechanisms and bureaucratic disaster assistance processes is the fear of local government corruption. This type of corruption has a doubly harmful effect, as it slows down the provision of federal aid, increases taxpayer costs, and also reduces the funds going directly to the impacted population—diverting those funds to corporate profits and to the direct beneficiaries of government corruption. In some cases, such as in Hurricane Katrina, the increased fear of public corruption resulting from a long history of convictions of government officials (the highest in the United States and two times higher than the second highest state), led to increased pressures from congress for accountability and further slowdowns of services and resource delivery (Voigt and Thornton 2015). As this example illustrates, the corruption itself, along with the history of corruption, had direct negative impacts on the disaster victims.

International research, by Nikolova and Marinov (2017), has shown a correlation between financial windfalls, such as an influx of disaster aid, and increased local government corruption. In the US context, research has also shown that major disasters create an environment in which public sector corruption is likely to increase as the temptation to misappropriate federal funds rises (Escaleras and Register 2016). Examples of corruption can include no-bid contracts, funds diverted to undamaged facilities, and payments where no work was performed. As these examples indicate, there is a clear connection between government corruption and corporate contracts. However, the assumption of corporate deservedness limits the efforts to prevent and identify this kind of corruption. This failure to fully address public corruption leads to increased death rates and damages as a direct result of the misappropriation of resources, a clear violation of the principle of distributive justice (Escaleras and Register 2016).

2.6 Deserving Victims, Post-Disaster Fraud, and Justice

As this chapter illustrates, the framing of victims as undeserving unless proven otherwise impedes Just Recovery and the creation of resilience. It places the onus of proof upon the impacted population while not requiring corporations or government to prove that they are behaving in a fashion that promotes justice and equity; in other words, the burden of proof is upon the victims in violation of the concept of principle of prima facie political equality (PPFPE). Revisiting the four principles identified in the Introduction as being the foundation of Just Recovery, current practices around proving deservedness violate the majority of these principles: they do not render governmental help accessible, do not provide safe housing for all, do not ensure access to resources for displaced persons, and do not ensure that there is fair and prompt assistance.

Notes

1. A more robust description of the various programs and policies will be provided in Chap. 4.
2. These biases are seen in media portrayals of protest as well, with people of color advocating for justice in the face of events such as police violence or environmental injustice (such as the indigenous protests of pipelines), portrayed as criminals, and their actions shown as meriting a militarized response.
3. Documentation of ownership is required at the time of the damage inspection, a necessary step in order to access FEMA disaster assistance (FEMA 2018).
4. An additional challenge was in the ability to verify identity, which is typically done through social security numbers. When social security numbers cannot be verified via an automated public records request, additional documentation is required (FEMA 2018).
5. Although many Latino workers did not experience harms directly from the hurricane, they did experience several adverse impacts during their role in disaster recovery, including long-term health impacts. It can therefore be argued that they constitute disaster victims as well.

References

Adams, V. (2013). *Markets of Sorrow, Labors of Faith: New Orleans in the Wake of Katrina*. Durham, NC: Duke University Press.

Brunsma, D., Overfelt, D., Picou, S. J., Bankston, C. L., III, Barnshaw, J., Bevc, C., et al. (2010). *The Sociology of Katrina: Perspectives on a Modern Catastrophe*. Lanham, MD: Rowman & Littlefield Publishers.

Calhoun, C. (2006). The Privatization of Risk. *Public Culture, 18*(2), 257–263.

Cuthbertson, B., & Nigg, J. (1987). Technological Disaster and the Nontherapeutic Community: "A Question of True Victimization". *Environment and Behavior, 19*(4), 462–483.

Dupuy, B. (2017, December 14). Hurricane Harvey Hit Black People the Hardest but They Are Still Waiting for Aid. *Newsweek*.

Escaleras, M., & Register, C. (2016). Public Sector Corruption and Natural Hazards. *Public Finance Review, 44*(6), 746–768.

Federal Emergency Management Agency. (2018). *Individual Assistance Program and Policy Guide (IAPPG)*. FEMA (Guidance Document).

Fehr, E., & Gächter, S. (2002). Altruistic Punishment in Humans. *Nature, 415*(6868), 137–140.

Fine, J., Mareschal, P., Hersh, D., & Leach, K. (2016). Contracting, Performance Management, and Accountability: Political Symbolism Versus Good Governance. *Journal of Strategic Contracting and Negotiation, 2*(4), 294–312.

Flavelle, C., Malik, N. S., & Smith, M. (2017, October 26). FEMA Probings $300 Million No Bid Contract for Puerto Rico Grid. *Bloomberg*.

Florido, A. (2018, March 22). Feds and Puerto Rico Reach Deal Allowing Disaster Recovery Loans to Start Following. *NPR*.

Freudenburg, W. (1997). Contamination, Corrosion and the Social Order: An Overview. *Current Sociology, 45*(3), 19–39.

Frontline. (2016, May 24). Who Profits When Disaster Strikes?—PBS Documentary.

Gotham, K. F. (2010). Disaster, Inc.: Privatization, Marketization, and Post-Katrina Rebuilding. *Perspectives on Politics, 10*(3), 633–646.

Hammer, D. (2015, August 23). Examining Post-Katrina Road Home Program: "It's More Than the Money, It's the Hoops We Had to Jump Through to Do It." *The Advocate*.

Horowitz, A. (2014, November). Hurricane Betsy and the Politics of Disaster in New Orleans' Lower Ninth Ward, 1965–1967. *The Journal of Southern History, LXXX*(4), 893–934.

Ibañez, G. E., Khatchikian, N., Buck, C. A., Weisshaar, D. L., Abush-Kirsh, T., Lavizzo, E. A., et al. (2003). Qualitative Analysis of Social Support and Conflict Among Mexican and Mexican-American Disaster Survivors. *Journal of Community Psychology, 31*(1), 1–23.

Jerolleman, A. (2013). *The Privatization of Hazard Mitigation: A Case Study of the Creation and Implementation of a Federal Program*. University of New Orleans Theses and Dissertations. Paper 1692. Retrieved from http://scholarworks.uno.edu/td/1692.

Jerolleman, A., Hodges, R., & Belblidia, M. (Forthcoming). Preventing Fraud vs. Preventing Risk Reduction—Are We Focusing Too Much on Making Sure That People Don't Cheat? In J. Kushma (Ed.), *Exploiting the Opportunities for Adaptation and Innovation*. Cambridge, MA: Elsevier Press.

Laska, S., Howell, S., & Jerolleman, A. (2018). "Built-In" Structural Violence and Vulnerability: A Common Threat to Resilient Disaster Recovery. In M. J. Zakour, N. B. Mock, & P. Kadetz (Eds.), *Creating Katrina, Rebuilding Resilience: Lessons from New Orleans on Vulnerability and Resiliency* (pp. 99–130). Oxford: Butterworth-Heinemann.

Mayer, B., Running, K., & Bergstrand, K. (2015). Compensation and Community Corrosion: Perceived Inequalities, Social Comparisons, and Competition Following the Deepwater Horizon Oil Spill. *Sociological Forum, 30*(2), 369–390.

Nikolova, E., & Marinov, N. (2017). Do Public Fund Windfalls Increase Corruption? Evidence from a Natural Disaster. *Comparative Political Studies, 50*(11), 1455–1488.

Oxfam Research Report. (2018, March 16). *Far from Recovery: Puerto Rico Six Months after Hurricane Maria*. Oxfam Research Report.

Peacock, W. G., & Ragsdale, A. K. (2000). Social Systems, Ecological Networks and Disasters: Toward a Socio-Political Ecology of Disasters. In W. G. Peacock, B. H. Morrow, & H. Gladwin (Eds.), *Hurricane Andrew: Ethnicity, Gender, and the Sociology of Disasters* (pp. 20–35). New York: Routledge.

Picou, J. (2009). Katrina as a Natech Disaster: Toxic Contamination and Long-Term Risks for Residents of New Orleans. *Journal of Applied Social Science, 3*(2), 39–55.

Reid, M. (2013). Disasters and Social Inequalities. *Sociology Compass, 7*(11), 984–997.

Scott, S., McSpirit, S., Breheny, P., & Howell, B. (2012). The Long-Term Effects of a Coal Waste Disaster on Social Trust in Appalachian Kentucky. *Organization & Environment, 25*(4), 402–418.

Sorensen, J. S., & Horrow, H. R. (2018, June 5). How Corruption Slows Disaster Recovery. *The Conversation*.

Swim, J. K., & Bloodhart, B. (2018). The Intergroup Foundations of Climate Change Justice. *Group Processes and Intergroup Relations, 21*(3), 472–496.

Tierney, K., Bevc, C., & Kuligowski, E. (2006). Metaphors Matter: Disaster Myths, Media Frames, and Their Consequences in Hurricane Katrina. *The Annals of the American Academy of Political and Social Science, 604*, 57–81.

Trujillo-Pagan, N. (2012). Neoliberal Disasters and Racialisation: The Case of Post-Katrina Latino Labour. *Race and Class, 53*(4), 54–66.

United States, Office of the Inspector General, OIG-16-01-D. (2015, October 6). FEMA Faces Challenges in Verifying Applicant's' Insurance Policies for the Individuals and Households Program.

Voigt, L., & Thornton, W. E. (2015). Disaster-Related Human Rights Violations and Corruption: A 10-Year Review of Post-Hurricane Katrina New Orleans. *American Behavioral Scientists, 59*(10), 1292–1313.

Zakour, M. J., & Grogg, K. (2018). Three Centuries in the Making: Hurricane Katrina from an Historical Perspective. In M. J. Zakour, N. B. Mock, & P. Kadetz (Eds.), *Creating Katrina, Rebuilding Resilience: Lessons from New Orleans on Vulnerability and Resiliency* (pp. 159–192). Oxford, UK: Butterworth-Heinemann.

CHAPTER 3

Survivor Agency

Abstract This chapter builds upon the notion of deservedness to explore its impacts upon the ability to exercise individual agency, as defined by the social work theory. The chapter explores the ways in which current post-disaster policies reduce individual agency, leading to exacerbated trauma and poor post-disaster outcomes, including negative impacts on social capital. These policies perpetuate structural violence on individuals, families, and communities. Furthermore, the loss of individual agency, coupled with the disparate outcomes that arise from complex processes and from the factors described in the previous chapters, further leads to the creation of a corrosive community in which families are pitted against each other and unable to work together for systemic change or to rely upon the social capital that might otherwise offer mutual assistance and aid.

Keywords Social capital • Individual agency • Collective efficacy • Corrosive community

This chapter applies the concept of individual agency, as defined by the social work theory and psychology, to the context of disaster recovery. It provides an exploration of the ways in which current post-disaster policies reduce individual agency, leading to exacerbated trauma and poor post-disaster outcomes. These policies are, at times, based on the notion that individuals will make decisions that are opposed to their best interests but are more often focused simply on reducing federal expenditures in the face of political pressures. Although there is clear research indicating that individuals do act according to what they believe to be in their best interests, the framing of individuals as poor decision-makers is often utilized and is connected back to the concept of the deserving victim (Kunreuther and Pauly 2005).

The intentional, and unintentional, reduction of individual agency through the administration of policies perpetuates structural violence on individuals, families, and communities, creating a secondary disaster. Furthermore, the loss of individual agency, coupled with disparate and varying levels of access to resources, further lead to the creation of a corrosive community in which families are pitted against each other and unable to work together for systemic change or to rely upon the social capital that might otherwise offer mutual assistance and aid (Aldrich 2017).

3.1 Survivor Agency

The definition of individual agency originates in the field of psychology and is inextricably connected to the concept of well-being, which is foundational to Just Recovery. Individual agency entails having full control over one's actions, including control over the determination of the purpose of the actions and the degree of effort to be expended—this is a key element of a capacities justice approach. In the context of disaster recovery, agency is fully connected to an individual's ability to participate in public deliberations and to make key decisions regarding their own recovery while expending reasonable levels of effort. These decisions might include taking actions to reduce future risks, making investments in recovery, and the more fundamental decision of whether or not to return. Governmental disregard for, or indifference to, individual agency reduces the resilience of individuals and their communities by limiting their options and forcing them toward particular outcomes (Laska et al. 2018). A clear example of this is found in the Lower Ninth Ward, following Hurricane Katrina, where homes built from mold resistant cypress, and already elevated, were declared over 50% damaged, and where roofs were declared 100% damaged, resulting in the denial of blue tarps, which further damaged the homes. Challenging these determinations required time and, often, the ability to travel to New Orleans in order to navigate a complex process. As a result, many homes that could have been repaired experienced further damages and homeowners were left without the option of returning (Allen 2007).

The need to prove deservedness, and the prioritization of neoliberal agendas and interests described in the previous chapters, impedes the agency of individuals and constitutes structural violence. Laws and policies are not designed or implemented so as to avoid this violence, increasing the trauma experienced by those who have been impacted by a disaster. In fact, the re-traumatization at the hands of bureaucratic procedures could even be considered an example of administrative evil,[1] in which unjust acts and outcomes are perpetuated and excused through the use of normal and routine procedures, even when these procedures violate equal protections or due process (Roberts 2013). Groups accustomed to experiencing administrative evil and structural violence, such as the poor or minority and indigenous communities, experience disasters as a continuation of ongoing trauma, while those unaccustomed to it such as the lower middle-class are traumatized by their inability to exercise their expected level of agency (Laska et al. 2018).

The experience of interacting with government recovery systems[2] following disasters is often described as traumatic, slow, time consuming, and energy-sapping (Dart 2018). In one anecdote, shared by Mark (2018), a woman in Puerto Rico faced extensive delays as a result of a misspelling of her name on the aid application. As a result of this error, the application was sent to the Fraud Inspection Unit, adding an expected additional four months to her application, while she was waiting for assistance and housing. A lack of Spanish speakers at the call center further exacerbated the stress occasioned by this bureaucratic error.

This blind adherence to bureaucracy despite outcomes is more prevalent among security organizations, such as Department of Homeland Security (DHS) and Federal Emergency Management Agency (FEMA), where the frequent concern with extreme threats under time pressure, including the need for secrecy which limits public accountability, serves as justification for discrimination against those viewed as a potential threat. Agency cultures can promote these discriminatory effects when they echo or amplify social judgments regarding who or what constitutes a threat (Roberts 2013). The perception of certain disaster victims as lawless, coupled with fears of civil unrest, lead to the perception of minorities as a threat and not as a resource. As a result, agencies such as FEMA focus resources on planning for threats such as civil unrest following a nuclear disaster, with a focus on marital law and mass internment, and not on supporting and promoting individual agency (Roberts 2013).

Administrative evil also takes place when bureaucratic processes do not flex to address the particular needs of those most vulnerable or in unique situations. As Roberts (2013, p. 393) describes "... individuals can make judgments that have discriminatory effects by relying on existing procedures even though the situation confronted is somewhat different from the situation that existing procedures were designed to meet." In other words, the need to stick to consistent procedures can serve to mask unjust actions and excuse the failure to put the human rights of survivors first and foremost in disaster recovery. This has the very real effects of displacement, traumatization, and at times injury or death.

This is an exercise of bureaucratic discretion that diminishes individual agency by reducing the range of potential options available to survivors and failing to provide justice. For example, FEMA's decision to deny Governor Rosello's request that the Direct Lease program be made available to Puerto Ricans who had evacuated to the mainland was denied on the grounds that it was not prudent to offer such assistance outside of the directly impacted area. As a result, the Puerto Rican diaspora has had to rely on the Transitional Shelter Assistance Program, a housing program that relies on hotel vouchers and is only designed for brief use. Families that had been forced to evacuate to the mainland after Hurricane Maria have faced repeated relocations, persistent threats of losing their vouchers, and in some cases have ended up homeless. The Puerto Rican diaspora has been penalized repeatedly for leaving the Island, despite often being forced to do so by the lack of access to basic services, healthcare, jobs, or schools. Rental stipends, when provided, are based upon rental costs back on the Island and insufficient in places such as Florida that were already facing a public housing crisis. Non-profits and churches have had to step in and help people whose vouchers expired or where insufficient. In some cases, families had vouchers revoked after FEMA determined that their homes back on the Island are livable, despite the many other impediments to returning.

This pushback against the presence of the Puerto Rican diaspora on the US mainland is both an assault on individual agency and a reflection of negative perceptions of the diaspora as undeserving victims. As Laska et al. (2018, p. 101) point out, "When a government dismisses the survivor's right to have agency, or ignores how successful agency can be enhanced with the help of government and non-profit agencies, a survivor's agency might not be used to enhance the survivor's condition." Furthermore, the portrayal of the victims as opportunistic, such as when public officials made statements that the Puerto Rican diaspora should have remained on the Island and were capitalizing on the hurricane, reduces public empathy regarding the reduction in individual agency.

In some cases, victim blaming is utilized to deflect attention from public failures, creating further post-disaster victimization (Phua 2008). This kind of victimization can occur at the hand of institutions, such as government agencies, or at the hands of other segments of victims. The concept of post-disaster victimization is closely connected to both survivor agency and to the processes described in the previous chapter. Examples of post-disaster victimization include the theft or diversion of resources intended for the victims; overcharging for goods and services which takes money away from victims; the exploitation of survivors looking for work; government failures in response; long delays from government or insurance companies; political persecution; and sullying the public image of victims (Phua 2008). Delays, in particular, have been shown to exacerbate trauma and in some cases have been purposefully utilized as punishment for the undeserving, something known as temporal domination (Laska et al. 2018).

Minority, indigenous, and immigrant populations are more frequently denied agency through purposeful exclusion in some cases and carelessness in others. Following Hurricane Harvey, immigrants were more likely to be in tenuous financial and social circumstances, less likely to have insurance or apply for disaster assistance, and more likely to need assistance obtaining additional medical care. This inability to meet basic needs imposes a severe limitation on individual agency as it prevents free and informed choice, free from any coercion (Wu et al. 2018).

3.2 Collective Efficacy and Social Capital

Negative impacts on individual agency, including the creation of a corrosive community, have direct and immediate impacts on the collective efficacy of a community and on its members' ability to exercise social capital, already suppressed by social distress (Mathbor 2007). Spokane et al. (2012) define collective efficacy as the extent to which community members will defend or stand up for each other in a crisis. They argue that the characteristics of temporary housing, which often place individuals far away from their communities or in group sites (particularly renters), serve to negatively impact recovery and reduce collective efficacy. Group sites, which are often utilized,[3] tend to lack community gathering spaces and create a disconnect between physical and social restoration, limiting the exercise of individual agency (Spokane et al. 2012). Other researchers have also found a connection between geographic connectivity and the ability to access kinship networks, which have been shown to promote resilience in the face of trauma. Social networks have been found to mitigate the negative experiences of evacuees, and in the case of the Vietnamese communities in New Orleans East following Hurricane Katrina, clearly facilitated the speedy return of the community when compared to similarly impacted communities (Li et al. 2010). However, these networks cannot operate as effectively when the entire network is impacted or when there is displacement (Godwin et al. 2013)

Researchers have described the emergence of therapeutic communities, in which altruistic norms and support emerge following a disaster and even among strangers (Bolin and Bolton 1986). The emergence of these support networks can assist individuals with retaining some measure of agency and provide assistance with the immediate demands of recovery. However, there are several constraints on family recovery, including the physical impacts of a disaster, the time required for recovery, and the mental health problems that can arise from feeling victimized and from having the familiar or normal disrupted. Social networks, kinship networks, and the therapeutic community are all strained over time and being to erode. Although this erosion can occur in different ways, over different time scales, some research has indicated that the relationship between a host family and an evacuee family begins to break down after one month (Bolin and Bolton 1986). This speaks to the need to promote policies that further individual agency, allowing for a wider range of options and reducing the strain on social networks while also speeding up the process of accessing recovery mechanisms.

Promotion of these policies requires full participation in recovery processes, particularly planning efforts, which can be both extremely limited when individual agency is constrained and can be limiting of agency when the possible futures for a community are determined by outsiders. Although planners have an ethical mandate to consider equity and social justice, most mainstream planning conversations do not incorporate the lived experience of individuals and do not take into account the experience of oppression. The displacement that often occurs following disasters, particularly of the poor and minority communities, further impacts their ability to participate. Understanding the lived experience of oppression, including its impacts on agency and perpetuation of participatory injustices, is critical to fully understanding how the vulnerable experience disasters and to then centering planning around community knowledge (Jacobs 2018). Planners may not be able to directly impact existing planning structures, but they can often influence the conditions that allow for participation, particularly by addressing the exercise of political power (Forester 1989).

When planners do not apply a critical lens to the political and social contexts of planning, they can perpetuate structural violence and systems of oppression and exacerbate trauma. This was the case with the Bring New Orleans Back Commission, which put forth a plan identifying neighborhoods as being targeted for green space and requiring those neighborhoods to show via engagement that they should be spared. This forced "engagement" at a time in which community members were still displaced and struggling to recover simply exacerbated trauma and did not truly allow their voices to be hard (Laska et al. 2018). This is an example of the need for both empathy and consideration of survivor agency.

A more recent example of exclusion from planning processes can be found in the creation of the Action Plan for the expenditure of Community Development Block Grant Disaster Recovery funds in Puerto Rico. Advocates have stated that the Action Plan was created without an adequate opportunity for community input. The comment period was limited to 14 days, with meetings held during the work day, limiting the ability of the average resident to attend. The meeting notices did not fully explain the importance of the process, and social media was utilized for advertising despite the lack of Internet and electricity among the general population. As a result, participants were mostly government officials and proposals presented included diverting money away from low-income housing. A coalition of over 80 advocates requested a 30-day extension but was denied (McLean 2018). This is a clear example of a public planning process reducing agency.

3.3 Reclaiming Agency

Recently, some researchers have looked at the role of collective narratives in supporting disaster recovery. These researchers have looked at both the role of social capital in the creation of narratives that facilitate individual agency and collective efficacy, as well as the ways in which residents can perform narratives as an organizing tactic through mechanisms such as tourism (Pezzullo 2010; Chamlee-Wright and Storr 2011). Pezzullo (2010) argues that tourism does not have to be depoliticized and that memories can be performed to nurture communities and foster collective recovery. This includes reclaiming agency through the use of tours as lobbying, organizing tactics, and generators of publicity. This purposeful utilization of narratives can serve to foster a collective identity of agency, strength, and survival. This stands in sharp contrast to the exploitative function that disaster tourism can often play.

This type of collective narrative was present in St. Bernard Parish following Hurricane Katrina where the community narrative was one of self-reliance and hard work. This perception prevented feelings of powerlessness reported by other communities and directly impacted the strategies for recovery that were selected by residents. People in St. Bernard reported returning because their networks could not be replicated elsewhere and they adopted a strategy of self-reliance without waiting for external assistance. The working class nature of the community treated the physical work of recovery as a challenge and not as something to be avoided (Chamlee-Wright and Storr 2011). However, it is important to note that St. Bernard Parish was a majority white community with some access to resources.

3.4 Survivor Agency and Justice

In cases where survivor agency is limited, or eliminated, by disaster policies and programs, community outcomes suffer. Agency is a key component of well-being and a fundamental requirement for Just Recovery, constituting the basis for one of the four key principles described earlier. Limitations upon individual agency also prevent equal access to resources and programs, and eliminate the possibility that adaptive capacity can be harnessed for risk reduction. Current disaster policies, including the challenges around their implementation, will be further discussed in the remaining chapters.

Notes

1. This term may appear unnecessarily harsh and is often utilized in the context of genocide and other horrendous historic events in which the government was complicit. However, fundamentally, the term implies the perpetuation of outcomes such as death through blind adherence to procedure. I would argue that this blind adherence to bureaucracy and procedure, in the face of escalating deaths, was clearly seen after Hurricanes Katrina and Maria. This is not to diminish the efforts of those government

employees who did exercise discretion to assist survivors. Furthermore, events such as the water crisis in Flint Michigan, where decisions made to save money despite public health consequences will lead to a lifetime of impacts for children exposed to lead, show that administrative evil does occur in the United States.
2. It is not the intent of this chapter to paint all government employees with the broad brush of indifference. Bureaucratic discretion is exercised daily by federal employees—including in support of disaster victims. Federal employees engaged in disaster work often commit to working long hours, far from home, for months at a time, and do so out of a desire to help the impacted citizens. However, that does not mean that all victims are seen as equally deserving and does not mean that agency cultures do not privilege cost savings over just outcomes.
3. After Hurricane Katrina, for example, over 70% of renters were placed in group sites. Those who remained in those sites the longest were those who were low income, elderly, or disabled (Spokane et al. 2012).

References

Aldrich, D. P. (2017). The Importance of Social Capital in Building Community Resilience. In W. Yan & W. Galloway (Eds.), *Rethinking Resilience, Adaptation and Transformation in a Time of Change*. Switzerland: Springer.

Allen, B. C. (2007). Environmental Justice and Expert Knowledge in the Wake of a Disaster. *Social Studies of Science, 37*(1), 103–110.

Bolin, R., & Bolton, P. (1986). *Race, Religion, and Ethnicity in Disaster Recovery*. Program on Environment and Behavioral Science, University of Colorado.

Chamlee-Wright, E., & Storr, V. H. (2011). Social Capital as Collective Narratives and Post-Disaster Community Recovery. *The Sociological Review, 59*(2), 256–274.

Dart, T. (2018, February 11). 'We've Been Forgotten': Hurricane Harvey and the Long Path to Recovery. *The Guardian*.

Forester, J. (1989). *Planning in the Face of Power*. Berkeley, CA: University of California Press.

Godwin, E., Foster, V. A., & Keefe, E. P. (2013). Hurricane Katrina Families: Social Class and the Family in Trauma Recovery. *The Family Journal: Counseling and Therapy for Couples and Families, 21*(1), 15–27.

Jacobs, F. (2018). Black Feminism and Radical Planning New Direction for Disaster Planning Research. *Planning Theory, 00*(0), 1–16.

Kunreuther, H., & Pauly, M. (2005). Insurance Decision-Making and Market Behavior. *Foundations and Trends in Microeconomics, 1*, 63–127. https://doi.org/10.1561/0700000002.

Laska, S., Howell, S., & Jerolleman, A. (2018). "Built-In" Structural Violence and Vulnerability: A Common Threat to Resilient Disaster Recovery. In M. J. Zakour, N. B. Mock, & P. Kadetz (Eds.), *Creating Katrina, Rebuilding Resilience: Lessons from New Orleans on Vulnerability and Resiliency* (pp. 99–130). Oxford: Butterworth-Heinemann.

Li, W., Airriess, C. A., Chen, A., Leong, K. J., & Keith, V. (2010). Katrina and Migration: Evacuation and Return by African Americans and Vietnamese Americans in an Eastern New Orleans Suburb. *The Professional Geographer, 62*(1), 103–118. https://doi.org/10.1080/00330120903404934.

Mark, J. (2018, April 23). The Case for Climate Reparations. *Sierra Club Magazine*.

Mathbor, G. M. (2007). Enhancement of Community Preparedness for Natural Disasters: The Role of Social Work in Building Social Capital for Sustainable Disaster Relief and Management. *International Social Work, 50*(3), 357–369.

McLean, D. (2018, August 25). Advocates Say Vulnerable Puerto Ricans Left Out of $1.5 Billion Recovery Planning Proceeds. *ThinkProgress*.

Pezzullo, P. C. (2010). Tourists and/as Disasters: Rebuilding, Remembering, and Responsibility in New Orleans. *Tourist Studies, 9*(1), 23–41.

Phua, K.-L. (2008). Post-Disaster Victimization How Survivors of Disasters Can Continue to Suffer After the Event Is Over. *New Solutions, 18*(2), 221–231.

Roberts, P. S. (2013). Discrimination in a Disaster Agency's Security Culture. *Administrative and Society, 45*(4), 387–419.

Spokane, A. R., Mori, Y., & Martine, F. (2012). Housing Arrays Following Disasters: Social Vulnerability Considerations in Designing Transitional Communities. *Environment and Behavior, 45*(7), 887–911.

Wu, B., Hamel, L., Brodie, M., Sim, S.-C., & Marks, E. (2018). *Hurricane Harvey: The Experiences of Immigrants Living in the Texas Gulf Coast*. Retrieved from https://www.kff.org/report-section/hurricane-harvey-the-experiences-of-immigrants-living-in-the-texas-gulf-coast-section-1/.

CHAPTER 4

Public Policy and Legislation

Abstract This chapter provides a deeper look at the current federal public policy and legislation surrounding disasters in the United States, including the role of federalism in its implementation at the state level. These policies are primarily aimed at reducing losses to the federal coffers and not at rebuilding communities; as a result, they privilege the recovery of homeowning nuclear families above all others. This has been seen in multiple disasters, including in the Road Home following Hurricane Katrina, where aid mechanisms were primarily geared toward homeowners despite the high percentage of renters and families in public housing. The chapter also provides a brief comparison of US policies and frameworks with international models that account for more attention to livelihoods and other structural considerations.

Keywords Federalism • Stafford Act • Emergency Management • Policy implementation • Administrative evil • Bureaucratic discretion

This chapter takes a closer look at the current federal public policy and legislation surrounding disasters in the United States, including the role of federalism in the implementation of these policies at the state level. National policies are primarily aimed at reducing losses to the federal coffers and not at rebuilding communities; as a result, they privilege the recovery of homeowning nuclear families above all others. The focus on homeowners is due to many factors, including the role of the National Flood Insurance Program (NFIP) and the strong property rights and homeownership culture within the United States. This has been seen in multiple disasters, including in the Road Home following Hurricane Katrina, where aid mechanisms were primarily geared toward homeowners despite the high percentage of renters and families in public housing (Adams 2013).

A full overview of the history of federal disaster policy is beyond the scope of this chapter but can be found in Jerolleman and Kiefer (2012). The chapter also provides a brief comparison of US policies and frameworks with international models, including those outlined in the Sendai and Hyogo Frameworks,[1] which account for more attention to livelihoods and other structural considerations.

4.1 US Disaster Policy

US disaster policies have primarily resulted from bursts of legislative action following major disasters; recent examples include Hurricane Katrina and Superstorm Sandy. As a result, the policies that have been created are fragmented and the majority of programs only provide funding following a presidentially declared disaster (Stehr 2006). From the late nineteenth century to the mid-twentieth century, very limited federal disaster assistance was available, but this changed following the Mississippi flood of 1927 when the federal government expanded its role in flood control by enacting the Lower Mississippi River Flood Control Act. From the mid-twentieth century, with the Federal Disaster Relief Act, the federal role in disasters was greatly expanded, and in 1956, the NFIP Act provided some degree of protection from flood losses. Between the 1950s and 1970s, federal involvement in disaster recovery increased to the extent that the federal share of disaster costs rose from 1% to 70%. This resulted in the passage of several pieces of legislation in response to increasing disaster expenditures, a pattern that continues to this day, while expenditures continue to rise.[2] One example of this is the Disaster Mitigation Act of 2000 (DMA2K), an amendment to the Robert T. Stafford Disaster Relief and Emergency Assistance Act (1988), which created a local hazard mitigation planning requirement. The Stafford Act, itself, had already created a requirement for state mitigation plans in order to promote conscious efforts at reducing disaster risk and future expenditures. The primary intent of DMA2K was to reduce the costs to the federal government from disasters through planning and local risk reduction projects (Jerolleman 2013). Unfortunately, research has shown that voters reward disaster relief spending but not spending on disaster preparedness or hazard mitigation (Healy and Malhotra 2008). Despite the Stafford Act and DMA2K, federal disaster spending tends to be far more focused on response and relief than on hazard mitigation, particularly hazard mitigation through programs such as the Pre Disaster Mitigation Grant Program that are dependent upon congressional appropriations outside of a declared disaster.[3]

4.2 Federalism

Public administration and emergency management scholars have long concerned themselves with the impacts of federalism on disaster policies, including response, within the United States. Under the US Constitution, public safety is a state responsibility, although state constitutions can enable local governments to take on rights and responsibilities relative to public safety.[4] Although the federal government does provide grants for equipment and training, as well as other types of assistance to local jurisdictions, this assistance always includes the state as an intermediary (Eisinger 2006). In fact, the majority of federal disaster assistance is administered by the states with local jurisdictions acting as sub-recipients, following a block grant model.[5]

The states are sovereign jurisdictions that establish the rights of local governments through their own constitutions. The level of state intervention in local decision-making, such as land-use decision, varies from state to state (Col 2007). When used properly, the federalist system can provide a great deal of flexibility, allowing states and jurisdictions to leverage mutual aid agreements when their own capacity is overwhelmed and to rely on higher levels of government when needed. However, mega-disasters require this coordination to happen very quickly and effectively, a challenge when so many levels of government are involved and when that coordination must be fluid and able to adapt to a changing environment (Landry 2008). Rapid innovation is a challenge for any level of government, much more so when intergovernmental coordination is required.

This model of disaster assistance allows for greater state control over recovery but can also lead to very different outcomes from state to state, depending upon agency cultures, staffing levels, capacity, approaches to contracting, and the extent to which the state prioritizes justice over special interests. Administering these programs requires navigating the complex intergovernmental partnerships described above while all levels of government are adversely impacted by budget constraints, the pattern of increasing losses, and the frequency of events (Pew Charitable Trusts 2018). As the national policy discussion has shifted toward increasing the state's financial responsibly for recovery, it has become apparent that the exact contribution of state and local governments is not consistently measured or accounted for, resulting in an inability to fairly judge the extent to which the current contributions are, or are not, appropriate (Pew Charitable Trusts 2018).

This is a significant concern because the creation of national policies that base disaster allocation upon some measure of the deservedness of states run the risk of simply perpetuating the vulnerability of the populations on the less affluent states or penalizing states that suffered from a lack of capacity to engage with federal processes and can be reasonably assumed to also lack the capacity to take on a greater role in recovery. If disaster risk reduction and hazard mitigation investments are taken as indicators of local engagement and investment, as has been proposed regarding the concept of a Public Administration Deductible, then the processes and policies that already discriminate against impoverished and rural states (areas less likely to successfully justify federal mitigation investments under a benefit-cost model) will simply be perpetuated, with the blame placed squarely upon the states themselves.

The receipt and successful management of prior disaster funds is not an adequate metric for measuring commitment to disaster risk reduction. Research has shown that disaster assistance is motivated more by political considerations than by need and that there is a correlation between disaster declarations, which open the door to a great deal of hazard mitigation funding, and politics (Olasky 2006; Landry 2008). Given this correlation, any policies that reward prior expenditures on risk reduction are simply rewarding states for having the capacity and political clout to more readily access and manage hazard mitigation grants or those with sufficient infrastructure and building stock (in terms of monetary value) to obtain grants where benefit-cost ratios are utilized (Healy and Malhotra 2008).

An alternative metric that has been proposed is to look at planning and regulatory efforts to reduce hazard risk by preventing unsafe development. In the United States, land-use decisions occur primarily at the local level although state regulations and enabling legislation serve to define the extent of local control.[6] There is a great deal of differentiation from state to state, with some states allowing relatively unfettered development and others placing strict controls on development in hazardous areas. The fact that states control development decisions that can create or exacerbate risks, while the federal government provides a great deal of monetary assistance following disaster impacts exacerbated by those same policies and local decisions, creates a classic externality. In other words, the costs are born by one party, while the benefits are born by another (Buchanan and Stubblebine 1962). In fact, the ease of access to flood insurance and disaster assistance has been cited as an impediment to local efforts at promoting risk reduction, as there are few incentives for local governments that are dependent upon the tax base and looking to promote the interests of the growth machine.

This system of competing incentives and priorities across all levels of government creates additional challenges for successful hazard planning, disaster risk reduction, and even disaster recovery (Stehr 2006). An example of this can be found in Houston, where the reservoirs constructed after the Great Houston Flood of 1935, including large areas of land designed to be flooded during releases, resulted in increased construction and development, some within the areas of land designed to be flooded. The county allowed these areas to be subdivided and constructed upon, with limited requirements for flood risk disclosure. In many cases, the flood risk of the properties within the development was not fully disclosed to the homebuyers but was instead only indicated on the platt maps. Although there are federal requirements for the disclosure of flood risk, these only apply to properties within the Special Flood Hazard Area (SFHA), a designation that did not take the inundation into account. Therefore, flood insurance was not required and many took this to mean that there was no flood risk. More recently, speculators purchasing the flooded homes and looking to resell them are choosing to not disclose flood risk because it is not required (Romero 2018).

Regardless of the metric utilized to determine the quality and the extent of local investments in disaster risk reduction, there is no doubt that the current mechanisms utilized for tracking local disaster spending are inadequate. A recent report by Pew Charitable Trusts (2018) found that spending is not tracked comprehensively by states and that the level of spending from state to state is highly variable. For example, Delaware covers 93% of state disaster spending with state programs, while Wyoming covers 0%. Of the 23 states that responded, 5 had state-level programs that mirrored federal programs, including Public Assistance, Individual Assistance, and programs for Hazard Mitigation. States also take very different approaches to cost sharing with local jurisdictions. Of the respondents, 40% covered at least a portion of the local cost-share for the Public Assistance Grant Program. The report recommended that a standardized approach to data collection be adopted if federal policy changes are enacted to either change the threshold for disaster assistance, or reward mitigation investments, or create a Public Assistance deductible requiring more upfront spending (Pew Charitable Trusts 2018).

4.3 Federal Disaster Programs

Federal Emergency Management Agency (FEMA) provides several types of resources, including funding, across all phases of Emergency Management. The assistance provided to survivors following a declared disaster is offered through three primary programs: the Public Assistance (PA) program, the Individuals and Households Program (IHP), and the Hazard Mitigation Grant Program (HMGP). This is in addition to the NFIP, also administered from within FEMA, which provides proceeds to those carrying flood insurance. IHP, HMGP, and the NFIP deal directly with individuals, although the state government always plays a role in the provision of federal assistance; PA is directed toward communities and intended to support the recovery of public facilities, including some facilities operated by non-profit organizations. There are provisions for risk reduction built into all of these programs in some way, although their application is uneven and not always successful. In many cases, individuals and families are required to seek a loan from the Small Business Administration (SBA) as either a prerequisite or an alternative to a direct grant. This has been a recurring source of confusion and frustration, with some households taking on unwanted or unsustainable debt while others were permitted to apply directly for IHP (Leicht 2017). This demonstrates a clear violation of procedural justice, as well as a reduction in individual agency (Box 4.1).

Box 4.1 Public Assistance for Infrastructure
Case Author: Paula R. Buchanan

The Robert T. Stafford Disaster Relief and Emergency Assistance Act of 1988 is the primary federal law used to initiate the process of the FEMA to coordinate financial and physical aid and relief to communities after disaster occurs, including Public Assistance, which is used to repair damaged infrastructure. One limitation of the Act is that infrastructure cannot be improved,[7] nor can repairs be funded, where damages are a result of a failure to properly maintain the damaged infrastructure. This weakness became especially apparent during the relief and recovery efforts that occurred in Puerto Rico, a US territory, after Hurricane Maria hit the island in 2017.

The country's existing infrastructure was antiquated and insufficient to meet the needs of the Island. However, infrastructure repairs are unable to increase the resilience of the Island and unable to address vulnerabilities created by a history of underinvestment in infrastructure.

Another significant source of post-disaster funding is the Community Development Block Grant Disaster Recovery Assistance Program (CDBG-DR) established in 1974. CDBG-DR funds are dependent upon congressional appropriations of funds following a disaster. The amount of funding appropriated has varied widely with funding exceeding 16 billion following Hurricanes Katrina and Sandy. The exact appropriation, and its timing, can become very politicized with varying levels of assistance provided to states with greater and lesser political influence. One key recommendation for ensuring distributive justice is that Housing and Urban Development (HUD) Department be provided with standing authority for CDBG-DR, up to a set amount (Leicht 2017). The total expenditures in CDBG-DR can be quite substantial, but expenditures by FEMA are much higher, due in large part to the number of programs administered by FEMA. CDBG-DR funds can be utilized for large community projects but can also be used to meet the needs of individuals and households. Although CDBG-DR funds were designed to incorporate a Low to Moderate Income requirement, ensuring that some portion of the funds benefit the most vulnerable, the percentage required has varied from event to event (Leicht 2017).

Administering the various grant programs, and interacting with the wide range of federal agencies involved in disaster recovery, can prove very taxing to state resources. Many states elect to create case management programs, often with the assistance of contractors, in order to navigate the various sources of funding and corresponding regulations. These programs, such as the Road Home in Louisiana, are frequently criticized for delays, inefficiency, and complexity; however, there are many challenges inherent in the programs themselves, as well as with the timelines for these programs relative to the timing for SBA, FEMA, and NFIP assistance. The complexity often leads to the Duplication of Benefits issues described in Chap. 2—Deserving Victims and Post-Disaster Fraud. The need to simplify and streamline the assistance process has been acknowledged by FEMA in its 2018–2022 Strategic Plan that includes the simplification of the assistance process as one of the three key goals for the agency (FEMA 2018). Leicht (2017) recommends the creation of one online disaster relief portal as an important first step toward streamlining the assistance process; one common application is not possible under privacy laws.

The difficulties inherent in the implementation of these programs will be discussed further in Chap. 5—Implementation Challenges. However, it is important to take a moment to discuss the inherent inability of current policies and regulations to ensure Just Recovery. Current disaster recovery programs and policies are designed primarily to compensate for measurable monetary losses, with no real consideration of need, resulting in higher awards for higher value properties and the perpetuation of existing inequalities (Kamel 2012). Homes located in neighborhoods that are experiencing declining property values are at times unable to receive sufficient funds for repairs and may also have been more likely to be without insurance or to lack the funds to cover deductibles and immediate repair needs. Deferred maintenance is also a recurring concern, as damage assessments are only intended to account for those losses directly attributed to the disaster event and not for any impacts determined to be due to the condition of the structure (Norton et al. 2018). The agency of impacted homeowners is routinely impacted by the complexity of government programs, all along different timelines, leading to a situation in which a full understanding of options and informed choice is simply not possible.

4.4 Human Rights and Disaster Recovery

Many recent calls for greater justice in disaster recovery, as well as in the policies surrounding disaster risk reduction, have called upon the US government to take more of a human rights-based approach. Generally speaking, the terminology utilized in emergency management within the United States has not drawn upon human rights framings. In fact, taking hazard mitigation as an example, the US concept of hazard mitigation is highly focused on the built environment and based primarily upon benefit-cost calculations that are unable to take justice into account. The concept of disaster risk reduction, more frequently used across Europe and by other nations, is defined by the United Nations Office for Disaster Risk Reduction (UNISDR) as including the components of preparedness and prevention, allowing for a consideration of a wider range of factors including as financial systems, food security, and underlying social vulnerability (Klima and Jerolleman 2014). According to the UNISDR website: "Disaster risk reduction is the concept and practice of reducing disaster risks through systemic efforts to analyze and reduce the causal factors of disasters. Reducing exposure to hazards, lessening vulnerability of people and property, wise management of land and the environment, and improving preparedness and early warning for adverse events are all examples of disaster risk reduction." This type of framing allows for greater consideration of the components of the V+ model than the framings contained within most US hazard policies and legislation.

The 2017 Global Platform for Disaster Risk Reduction reaffirmed the commitments of the Sendai Framework (which built upon the Hyogo Framework) and included among its priorities investing in disaster risk reduction for resilience, enhancing preparedness, and strengthening disaster risk governance. Specific commitment outlined within the framework include a consideration of cultural heritage of indigenous peoples, addressing underlying risk drivers, making resilience affordable, strengthen community resilience for the poor and vulnerable, and promoting a people-centered approach to development (UNISDR 2007, 2012).

The United States has frequently expressed verbal support for the Hyogo Framework and international efforts for its implementation. Public statements have explicitly referenced the connections between sustainable development and disaster risk reduction, although most statements have been outward-facing—with the United States discussing the Hyogo Framework in the context of US aid to other nations and not in terms of planning efforts within the United States. For example, a 2012 Interim National Progress Report on the Implementation of the Hyogo Framework for Action within the United States cited multiple examples of progress toward meeting the Hyogo Framework (USA 2012). These examples all focused on existing US efforts that on the surface do not appear to have been informed by the Hyogo Framework, but some of which, like the Whole Community Approach, can be interpreted to indicate a commitment to some of the components of the framework.

Looking more explicitly at human rights, and international human rights law, human rights standards are only enforceable in the United States when they are codified in local, state, or federal law. Although the United States was a leader in the creation of the Universal Declaration of Human Rights, domestic legal protections for human rights are not fully aligned with international standards. In some instances, the United States has directly opposed international human rights laws, such as when discriminatory practices such as Jim Crow laws were at risk of being challenged were an international treaty to be ratified (The Advocates for Human Rights n.d.).

4.5 Public Policy, Legislation, and Justice

Adopting a human rights-based approach to disaster policy, one in which the well-being of the impacted families is the goal, as opposed to simply financial restitution of eligible losses, would help to achieve the principles of a Just Recovery. Coupling this approach with improved coordination between policies, allowing for more informed decision-making and more equal access to resources is also needed. Generally speaking, disaster policy in the United States is fragmented and focused more on the reduction of costs to the federal government than on disaster risk reduction, and certainly not on just outcomes.

Notes

1. Although US policies and frameworks are country specific, and the international frameworks are sets of guiding principles to which nations sign on, the comparison is useful in highlighting the differences in the terminology utilized and the factors considered.
2. One of the primary reasons for rising disaster losses is the development patterns that are placing more housing in high-risk locations. This housing stock includes middle-class homes, as well as second homes in desirable locations such as by the ocean or by mountains.

3. The Disaster Recovery Reform Act of 2018, signed into law in October of 2018, is intended to direct more funding towards preparedness, but it is too soon to know how it will be implemented.
4. In this context, public safety is usually taken to mean police, fire, and other first-response functions. The federal government retains responsibility for the protection of borders and the military.
5. One exception to this was Project Impact, a program that directed funding to local communities for preparedness and risk reduction actions. However, the nature of the funding, including the fact that it bypassed the states, was one of the reasons that the program was eliminated.
6. There are examples of federal incentives for the adoption of building codes and stronger land-use practices. These include the Community Rating System and requirements built into CDBG-DR Action Plans.
7. It is, however, possible to include some mitigation measures as well as to meet codes and standards that may have become more stringent in the time elapsed since the original construction.

References

Adams, V. (2013). *Markets of Sorrow, Labors of Faith: New Orleans in the Wake of Katrina*. Durham, NC: Duke University Press.

Buchanan, J. M., & Stubblebine, W. C. (1962). Externality. *Economica, 29*(116), 371–384.

Col, J.-M. (2007, December). Managing Disasters: The Role of Local Government. *Public Administration Review*.

Eisinger, P. (2006, August). Imperfect Federalism: The Intergovernmental Partnership for Homeland Security. *Public Administration Review*.

Federal Emergency Management Agency (FEMA). (2018, July 12). *2017 Hurricane Season FEMA After-Action Report*.

Healy, A. J., & Malhotra, N. (2008). Preferring a Pound of Cure to an Ounce of Prevention: Voting, Natural Disaster, and Government Response. Retrieved from http://www.sscnet.ucla.edu/polisci/cpworkshop/papers/Healy.pdf.

Jerolleman, A. (2013). *The Privatization of Hazard Mitigation: A Case Study of the Creation and Implementation of a Federal Program*. University of New Orleans Theses and Dissertations. Paper 1692. Retrieved from http://scholarworks.uno.edu/td/1692.

Jerolleman, A., & Kiefer, J. (2012). *Natural Hazard Mitigation: A Casebook for Academics and Practitioners*. Boca Raton, FL: CRC Press.

Kamel, N. (2012). Social Marginalisation, Federal Assistance and Repopulation Patterns in the New Orleans Metropolitan Area Following Hurricane Katrina. *Urban Studies, 49*(14), 3211–3231.

Klima, K., & Jerolleman, A. (2014). Bridging the Gap: Hazard Mitigation in the Global Context. *Homeland Security & Emergency Management*.

Landry, M. (2008). Mega-Disasters and Federalism. *Public Administration Review*, Special Issue.
Leicht, H. M. (2017, July 19). *Rebuild the Plane Now: Recommendations for Improving Government's Approach to Disaster Recovery and Preparedness.* The Community Preservation Corporation.
Norton, R., MacClune, K., Venkateswaran, K., & Szönyi, M. (2018). *Houston and Hurricane Harvey: A Call to Action.* Zurich, Switzerland: Zurich Insurance Company Ltd.
Olasky, M. (2006). *The Politics of Disaster: Katrina, Big Government, and a New Strategy for Future Crisis.* Nashville, TN: W Publishing Group.
Pew Charitable Trusts. (2018, June). *What We Don't Know About State Spending on Natural Disasters Could Cost Us: Data Limitations, Their Implications for Policymaking, and Strategies for Improvement.* A Report.
Romero, S. (2018, March 23). Houston Speculators Make a Fast Buck from Storm's Misery. *The New York Times.* Retrieved from https://nyti.ms/2g6scex.
Stehr, S. D. (2006). The Political Economy of Urban Disaster Assistance. *Urban Colloquy Urban Affairs Review, 41*(4), 492–500.
The Advocates for Human Rights. (n.d.). Retrieved from https://www.theadvocatesforhumanrights.org/human_rights_and_the_united_states.
United Nations Office for Disaster Risk Reduction (UNISDR). (2007). Hyogo Framework for Action 2005–2015: Building the Resilience of Nations and Communities to Disasters. Retrieved from http://www.unisdr.org/we/inform/publications/1037.
United Nations Office for Disaster Risk Reduction (UNISDR). (2012). The 10 Essentials for Making Cities Resilient. Retrieved from http://www.unisdr.org/campaign/resilientcities/toolkit/essentials.
United States of America. (2012). Interim Progress Report of the Implementation of the Hyogo Framework for Action (2011–2013). Retrieved from http://www.preventionweb.net/files/28816_usa_NationalHFAprogress_2011-13.pdf.

CHAPTER 5

Implementation Challenges

Abstract This chapter builds upon the information provided in Chap. 4, with a focus on the actual implementation of policies and the resulting impacts of bureaucratic processes. The chapter argues that although the implementation of the current disaster policy and legislation has consistently led to the exacerbation of inequalities and unjust outcomes, there is a great deal that can be improved (or worsened) simply through the interpretation of the rules and regulations and the exercise of bureaucratic discretion—or perpetuation of administrative evil. The chapter describes some of the ways in which implementation practices and challenges lead to disparate outcomes, including the impacts of agency cultures that have become very focused on preventing excessive spending.

An additional implementation challenge stems from the reliance on private sector contractors, along with other partners, often utilizing contracting vehicles and other vehicles that create incentives for a focus on projects and applicants with more readily resolved cases.

Keywords Privatization • Stafford Act • Emergency Management • Policy implementation • Administrative evil • Bureaucratic discretion

The implementation of the current disaster policy and legislation has frequently led to the exacerbation of inequalities and unjust outcomes, an outcome that can be greatly improved (or worsened) simply through the interpretation of the rules and regulations and the exercise of bureaucratic discretion. This chapter describes some of the ways in which implementation practices and challenges lead to disparate outcomes, including the impacts of agency cultures that have become very focused on preventing excessive spending.

An additional implementation challenge stems from the reliance on private sector contractors, along with other partners, often utilizing contracting vehicles and other vehicles that create incentives for a focus on projects and applicants with more readily resolved cases. This was seen following Hurricane Katrina, with the Road Home, again following Superstorm Sandy, and continues to be seen through the selection of mitigation projects that are more readily fundable (Jerolleman 2013).

5.1 Program Implementation

Public administration scholars widely recognize the role that implementation plays in the success or failure of government programs and policies. The literature on program implementation, broadly speaking, consists of both a bottom-up and a top-down view. An important concept in the literature on bottom-up review is that of street-level bureaucracy which focuses on the role of discretion in service delivery at the street level (Lipsky 1980). This type of discretion impacts the lived experience of individuals interacting with the government and, over time, essentially becomes the policy.[1] It can be expressed through assistance in navigating bureaucratic hurdles, particularly when there are feelings of empathy or in-group dynamics at play; or, it can be expressed through administrative evil when individuals are seen as undeserving and rules and regulations are applied regardless of the needs of those interacting with the government. Alternatively, a top-down approach to the study of program implementation looks more at the role of implementing agencies, and authorizing legislation and statutes, relative to the problem being addressed through the government program (Sabatier 1986). Looking at implementation through this lens, success is achieved when the objectives of the legislation are achieved.

These two approaches to understanding the success or failure of implementation are not mutually exclusive and in fact can be combined to create a better understanding of why policy objectives so often fail. Public administration scholars repeatedly identify implementation challenges as the reason for the failure of public policies in arenas such as disasters, education, and healthcare. A particular challenge to implementation lies in the need to address local needs, including creating programs that are flexible and adaptable, through centralized efforts that can be applied throughout the United States. Excessive top-down control can render a program overly rigid and exclude the local actors who better understand regional and state particulars that impact implementation.

Additional challenges to implementation arise when there are multiple levels of government involved or when additional external actors become involved. This type of complexity is present in disaster recovery, where states administer federal funds and all levels of government both utilize private sector contracts and partner with other entities such as non-profits—in some cases—through similar contracting mechanisms.

A recent example of the challenges of implementing federal programs can be found in Texas, following Hurricane Harvey. Governor Abbott directed the General Land Office to manage short-term housing and emergency repair efforts, both of which have typically been administered more directly by Federal Emergency Management Agency (FEMA). This model of increased local control was touted as a new and more efficient alternative to a FEMA-led effort and as being in line with the national conversation regarding the appropriate role of the state and federal governments in disasters. However, the state offices had to learn regulations, negotiate contracts, and navigate the complexity of the FEMA programs. As a result, the effort has lagged behind federally led efforts in other areas. This may indicate that the increased frequency of engagement with these programs at the federal level allow for the creation of more institutional memory than is held at the state level where these interactions are less frequent. The first trailer was not in place until 43 days after Harvey, as compared to the placement of 900 trailers in the 3 weeks following Hurricane Katrina (Weissert and Schmall 2018). Of the 890,000 families that applied for assistance, 40,000 of which needed immediate short-term housing assistance, only 900 had received it after three months (Formby 2018). To be fair, Texas faced unique challenges with local land ordinances, but the challenge of navigating the complexity of a federal program while facing local realities is experienced by all states. Even the private sector, sometimes tapped as a more efficient mechanism of program delivery, faces delays and challenges in the face of such complexity (Jerolleman 2013).

5.2 Implementation of Existing Policies and Programs

As was previously described, current federal policies and programs around disasters allow for a certain measure of discretion and interpretation that street-level bureaucrats can elect to either exercise or not. This discretion is exercised in the context of the framings presented in Chap. 2—Deserving Victims and Post-Disaster Fraud—where some victims are seen as deserving of assistance and others are not. As a result, the implementation of policies and programs is uneven, often benefiting certain populations above others. As Chap. 4—Public Policy and Legislation—described, some of these uneven benefits have to do with the design and purpose of the programs, while others stem from challenges in implementation.

Although many of the programs utilized following a declared disaster are made available and funded via formulas set forth by legislation, other programs such as Community Development Block Grant Disaster Recovery Assistance Program (CDBG-DR) are dependent on congressional allocations following the disaster. This is also true for programs such as emergency food stamps that are typically issued to assist the survivors of a natural disaster in maintaining access to food, a basic human right, during their initial recovery. As a result, congress and the federal government exercise a certain amount of discretion where post-disaster assistance is concerned, some through democratic and legislative processes, and others through the setting of priorities and the framing of particular events and populations. Funding allocations reflect values and priorities, including the perception of the deservedness of the impacted communities. In all instances, the political power exercised by the impacted state, or territory, has an impact on its ability to advocate for such appropriations.

Pre-existing patterns of exclusion and limitations placed upon local sovereignty, such as the austerity measures imposed on Puerto Rico as an impoverished territory, play a role as well. Unlike the mainland, Puerto Rico has a capped annual budget for the food stamp program, one that already does not prove sufficient to meet the needs of all qualifying families. Emergency food stamps, and additional food stamp allocations, were not issued following Hurricane Maria. Several newspaper articles, reports, and other documents have drawn attention to the sharp contrast between the quality, speed, and scope of the federal response to Hurricane Harvey as opposed to that following Hurricane Maria. Although low-income neighborhoods in the Houston area were disproportionately impacted and remain uninhabitable, the recovery efforts in Puerto Rico reflect the disempowerment of local and state decision makers coupled with a loss of basic needs such as power and water (Deseret 2018). This will be discussed further in Chap. 6—Disaster Risk Reduction and Creation.

Another example of implementation decisions reflecting priorities, and of the exercise of administrative evil, can be seen in the demolition of homes in the Lower Ninth Ward following Hurricane Katrina. Data from February of 2006 through September of 2007 showed a far more intense pattern of demolitions in the Lower Ninth Ward, often without sufficient notice to residents who felt that their level of damages had been inflated and did not merit such speedy demolition. This pattern exceeded that in other heavily impacted neighborhoods, such as Lakeview, which were not left off rebuilding plans in the same way (Hatcher et al. 2012). The speed of demolitions did not allow residents to return and rebuild, instead limiting their range of options and forcing relocation.

5.3 Limitations of Current Programs and Policies

As the previous chapter explained, the vast majority of federal assistance is provided following a presidentially declared disaster, and the costs of such assistance have been rising. Much of the federal policy and legislation surrounding disaster risk reduction was created to reduce the costs of disasters, to the federal government, and as a result prioritizes assistance to homeowners and to those who carry insurance. This focus on homeownership reflects deeply held American values regarding land ownership while also privileging the recovery of the middle class, a group seen as more deserving of assistance. The inability of the current programs to meet the needs of renters following events such Hurricane Katrina (which damaged 33% of the rental housing stock in the state of Louisiana) and Superstorm Sandy has been clearly documented (Gotham 2010). Even when efforts have been made to support the rehab of rental properties in the impacted area, there have been problems with delays, limited participation, and almost no ability to ensure that adequate rental stock remains at an affordable rate (Kamel 2012). The unique challenges faced by renters, and by those in public housing, result in very disparate outcomes described in the following chapters.

Assumptions about household composition also mirror American values of the nuclear family, with multi-generational households excluded from receiving full aid and assistance (Sou and Aponte-Gonzalez 2017).

Low-income, minority, and indigenous communities often utilize alternative models for household composition, with multiple families sharing space or multiple generations co-habiting. Current application processes only permit one application per residence. These same communities are more likely to have utilized informal mechanisms for the transfer of property and may not have a clear title in place at the time of the event, another factor that has consistently limited access to disaster aid. Under the Road Home Program, created in Louisiana following Hurricane Katrina, the contract provisions between the state and the contractor allowed for payments based on the number of cases closed, creating a disincentive for the contractors to spend time on appeals and more difficult cases (Jerolleman 2013). A similar focus on resolving easier cases, often involving families that are not the most in need, has been identified following the Deepwater Horizon Oil Spill and other disasters where contractors are utilized as case managers.

There are several additional assumptions inherent in the current models for the provision of disaster assistance including that the victims of the disaster will have transportation to the Disaster Assistance Centers, will be able to spend extensive amounts of time on the process of recovery including time away from family and livelihood commitments, and will be able to qualify for and navigate the complex set of post-disaster programs. A great deal of prior research has found that access to government recovery programs is uneven, with minorities such as black and Latino populations frequently unable to qualify for assistance. Rural communities also face several disadvantages in attempting to access aid, as do the elderly and disabled (Bolin and Bolton 1986). Some of these challenges can be addressed through modifications to application processes and requirements, while others would require extensive changes to legislation and to the Code of Federal Regulations, in order to modify eligibility criteria and promote Just Recovery.

The complexity of the process, including that of the appeals process, creates a significant barrier to agency for many individuals. In and of itself, the mix of state and federal programs that are brought to bear and in some cases created on the fly following a disaster is confusing, CDBG-DR rules, for example, vary from appropriation to appropriation. Many families struggle to navigate the complicated bureaucracy and regulations even when language and literacy don't present additional barriers as is often the case. The process of applying for multiple different sources of funding requires the completion of multiple processes, such as making an insurance claim, applying for Individuals and Households Program (IHP) funding, and seeking additional services from local providers. In order to avoid Duplication of Benefits, applicants must often submit information multiple times, as the damage assessment process results in the receipt of insurance proceeds, or other sources of assistance are secured. The burden of documentation is placed solely upon the applicants, and successfully navigating it often requires appealing rejection letters that are mailed out as standard practice while applicants wait to see if they qualify for Small Business Administration (SBA) loans or other mechanisms (Vinik 2018).

Concerns regarding the damage assessment process have been raised in Texas following Hurricane Harvey, with some community advocates expressing concerns about the use of the FEMA damage assessments as the basis for the state's Draft Rebuilding Plan. They argue that the damage assessments fail to capture the full extent and scale of losses, causing particular damage to low-income victims. A coalition of advocates, led by Texas Housers, submitted the critique in comments regarding the plan but received a negative response from the Texas General Land Office. According to the Texas General Land Office, the federal damage thresholds are set by FEMA and failing to utilize them would lead to delays for all households seeking assistance (WBAY 2018). This type of response, alleging that efforts to increase just outcomes would cause delays across the board, is a common response from state and federal government officials when such concerns are raised. This type of response positions those requesting justice as obstructionists slowing down recovery for all by increasing the complexity of the process. Unfortunately, this detracts from any substantive efforts to promote change, as even those most adversely affected cannot afford to have longer wait times and bow under external pressures—often without knowing just how much additional delay, if any, would have been experienced.

The timelines over which assistance is provided and decisions are made regarding the different programs, including changes during the process of implementation, also prove exceptionally challenging for lower-income families. As Kamel (2012, p. 3216) noted, the "…combination of repeated changes in deadlines and eligibility criteria, long processing time and generally low grant amounts severely hindered recovery especially for low-income and minority residents." IHP, for example, is only provided for 18 months, a timeframe that is often unrealistic following a catastrophic event where families are waiting on federal funds or third-party decisions regarding rental housing. Families are often forced to spend some portion of funds on subsidence or may make some initial repairs without fully understanding the types of documentation required for reimbursement. Denials due to insufficient damage or deferred maintenance are also common for lower-income families, and these denials impact the level of identified need to which federal programs, such as CDBG-DR, respond. In other words, the allocations of CDBG-DR assistance are based initially on documented damages, as recognized by FEMA.

The impacts of this implementation process have impacted recovery for low-income neighborhoods impacted by Hurricane Harvey. According to Vinik (2018), residents of low-income neighborhoods impacted by Hurricane Harvey often lacked flood insurance due to affordability issues, lack the savings to make repairs not covered by FEMA, and either lack the income to qualify for an SBA loan or cannot take on the debt. As a result of this, in neighborhoods such as Kashmere Gardens, where two-thirds of the residents are black and the median income is only $23,000, many were denied FEMA assistance due to their inability to carry flood insurance following the receipt of prior disaster assistance.

Another example of the creation and perpetuation of inequality through the process of implementing recovery programs can be found in the creation of the Gulf Opportunity Zone and the use of CDBG-DR funds following Hurricane Katrina. The Go Zone provided low-interest bonds and tax incentives to an area covering 60,000 square miles across Mississippi, Alabama, and Louisiana, and including undamaged areas. The majority of the funds went to business in the least damaged areas, as they were more able to access the incentives in a shorter time frame. Large businesses, such as petrochemical companies, were also very successful at navigating these regulations and accessing the tax credits—despite their role in the creation of risk. In New Orleans, none of the bonds that were granted were utilized to redevelop in the most damaged areas. As this example shows, the implementation of the program gave certain advantages to the least impacted businesses, arguably those that least needed the incentives in order to ensure their survival. A second example is the expenditure of CDBG-DR funds in Louisiana, where the Road Home program utilized a non-racial housing valuation system, one that ignored the disparities in pre-storm values that disproportionately affected African American neighborhoods and resulted in insufficient grants for home repairs in those communities. In this particular instance, the Greater New Orleans Fair Housing Alliance sued over discrimination against African American neighborhoods and a settlement was reached in 2011 that provided additional CDBG funds for applicants whose grants had been based solely on pre-storm values and not on the cost of repairs (Gotham 2010). As with many policies and programs, the metric of success was considered an expenditure of the funds and the granting of incentives, not distributive or participatory justice (Gotham 2010). The legacy of the Road Home was the perpetuation and reinforcing of racial and class housing segregation because these long-standing patterns of segregation were ignored in program design.

5.4 Using Current Policies and Regulations to Promote Just Outcomes

A great deal has been written, here and elsewhere, regarding the limitations of current policies and regulations. However, current regulations do include provisions that, when enforced, can serve to require more just outcomes. For example, the civil rights and fair housing requirements tied to CDBG-DR funding are not historically enforced, creating a missed opportunity to reduce segregation and disinvestment in communities (Sloan and Fowler 2015). In a few instances, often following legal action, the Fair Housing Act of 1968 has been successfully utilized to increase the fairness and effectiveness of CDBG-DR programs. The Fair Housing Act requires the furthering of fair housing, not just refraining from discrimination, which following a disaster includes preventing the permanent displacement of protected groups and ensuring that there is investment in infrastructure and economic development for areas that have historically been disadvantaged (Sloan and Fowler 2015). One of the successful utilizations of the Fair Housing Act is the successful suit in New Jersey in 2014 that required additional funds for renters whose applications for assistance had been erroneously rejected, with twice the rejection rate for African Americans as for others. Similarly, in St. Bernard Parish, following Hurricane Katrina, the Fair Housing Act was used to block discriminatory housing ordinances that were attempting to block multifamily housing.

5.5 Implementation Challenges and Justice

One key recommendation to promote just outcomes is to assess damage and funding needs for recovery, relative to income, and to not simply set a minimum dollar damage threshold (Vinik 2018). This type of approach would privilege the distribution of goods relative to need, promoting well-being, and allowing greater agency for recovery. However, a fully just model must also take into account renters and those whose needs cannot be covered through existing frameworks. Therefore, an additional recommendation that an institutionalized system be developed to evaluate the CDBG-DR Action Plan for civil rights compliance is also necessary (Sloan and Fowler 2015).

Robust community planning can serve to enhance participatory access, particularly where decisions regarding allocations of assistance and future development are concerned, but this type of planning for recovery can be impacted by trauma, the strain on social networks, and the destruction of infrastructure (Reardon et al. 2009).

Current implementation practices are not in line with a capacities justice framework, nor do they follow the four principles put forth as the basis for Just Recovery. Programs would be far more effective if their goals and implementation strategies included a focus on enhancing equity through recovery and an understanding of underlying socio-spatial inequalities (Gotham 2010).

Note

1. The on-the-ground impacts of policies, and the interactions of individuals with government, stem directly from the decisions made by street-level bureaucrats.

References

Bolin, R., & Bolton, P. (1986). *Race, Religion, and Ethnicity in Disaster Recovery.* Program on Environment and Behavioral Science, University of Colorado.

Deseret News Editorial Board. (2018, June 8). In Our Opinion Disaster Relief Can Sometimes Be a Disaster. Retrieved from https://www.deseretnews.com/article/900020973/in-our-opinion-disaster-relief-can-sometimes-be-a-disaster.html.

Formby, B. (2018, February 27). Abbot and FEMA Are Using Harvey to Reinvent Disaster Response. *The Texas Tribune.*

Gotham, K. F. (2010). Disaster, Inc.: Privatization, Marketization, and Post-Katrina Rebuilding. *Perspectives on Politics, 10*(3), 633–646.

Hatcher, L. J., Strother, L., Burnside, R., & Hughes, D. (2012). The USACE and Post-Katrina New Orleans: Demolition and Disaster Clean-Up. *Journal of Applied Social Science, 6*(2), 176–190.

Jerolleman, A. (2013). *The Privatization of Hazard Mitigation: A Case Study of the Creation and Implementation of a Federal Program.* University of New Orleans Theses and Dissertations. Paper 1692. Retrieved from http://scholarworks.uno.edu/td/1692.

Kamel, N. (2012). Social Marginalisation, Federal Assistance and Repopulation Patterns in the New Orleans Metropolitan Area following Hurricane Katrina. *Urban Studies, 49*(14), 3211–3231.

Lipsky, M. (1980). *Street-Level Bureaucracy: Dilemmas of the Individual in Public Services.* New York: Russell Sage Foundation.

Reardon, K. M., Green, R., Bates, L. K., & Kiely, R. C. (2009). Overcoming the Challenges of Post-Disaster Planning in New Orleans: Form the ACORN Housing University Collaborative. *Journal of Planning Education and Research, 28*, 391–400.

Sabatier, P. A. (1986). Top-Down and Bottom-Up Approaches to Implementation Research: A Critical Analysis and Suggested Synthesis. *Journal of Public Policy, 6*(1), 21–48.

Sloan, M., & Fowler, D. (2015). *Lessons from Texas: 10 Years of Disaster Recovery Examined.* Texas Appleseed.

Sou, G., & Aponte-Gonzalez, F. (2017, December). *Making Efforts Count After Irma and Maria: Household Relief and Recovery in Puerto Rico.* University of Manchester, UK: Policy Brief.

Vinik, D. (2018, May 29). 'People Just Give Up': Low-Income Hurricane Victims Slam Federal Relief Programs. Politico.

WBAY. (2018, April 27). Hurricane Harvey Recovery Funds May Prioritize Wealthy, Advocates Say. Retrieved from https://www.wbay.com/content/news/Harvey-recovery-funds-may-prioritize-wealthy-advocates-say-481089261.html.

Weissert, W., & Schmall, E. (2018, February 22). Texas' Vow to Streamline Harvey Recovery and Aid Backfires. The Washington Post.

CHAPTER 6

Disaster Risk Reduction and Creation

Abstract This chapter delves more deeply into hazard mitigation and disaster risk reduction, both of which are intended to take place prior to an event but most often do not. Among other topics, this chapter discusses the ways in which policies geared at disaster risk reduction can simply lead to gentrification, as only those who can afford the cost of reduced risk may remain and as state and local governments elect to prioritize economic development above resilience and risk reduction. The chapter also describes the ways in which risk is exacerbated for those who are displaced in the process.

Keywords Disaster risk reduction • Disaster risk creation • Climate adaptation • Hazard mitigation • Climate gentrification

This chapter delves more deeply into hazard mitigation and disaster risk reduction, both of which are intended to take place prior to an event but most often do not. In many instances, the policies geared at disaster risk reduction can simply lead to gentrification, as only those who can afford the cost of reduced risk may remain and as state and local governments elect to prioritize economic development above resilience and risk reduction (GAO 2014; NRC 2012). The chapter also describes the ways in which risk is exacerbated for those who are displaced in the process.

6.1 Disaster Risk Creation

The Vulnerability-Plus (V+) model described at the start of this text includes, as one of its underlying assumptions, the fact that vulnerability is not distributed equally between communities and people. The spatial distribution of vulnerability, created over time as a result of social processes, has served to limit access to capitals and create increased disaster risk. The historic creation and exacerbation of disaster vulnerability have led to the current hazard landscape. Ignoring this history, through attempts at colorblind climate adaptation will only serve to perpetuate prior injustices (Hardy et al. 2017). Although many historic processes have created landscapes of risk, including the forced removal of indigenous populations from their lands to reservations,[1] redlining of minority areas, destruction of communities for the creation of the highway system, and rising income inequality, this chapter will only focus on a few of these processes as examples of the creation of disaster risk. More recent examples from Hurricane Katrina, as well as other events, are also utilized throughout this chapter.

The practice of redlining by the Home Owner's Loan Corporation (HOLC) in the 1930s, in which neighborhoods were identified as being high-risk and therefore not suited for investment, created much of the racial and economic segregation that exists to this day (Mitchell and Franco n.d.). Neighborhoods identified as high-risk provided much fewer opportunities for homeownership or other types of loans, concentrating minorities in neighborhoods where little economic investment was taking place. This history of limiting access to credit and loans continues to impact residents to this date, through the persistence of poverty and chronic challenges, such as the lack of a clear title that limits access to disaster relief. Of the communities identified as high-risk, 74% remain classified as Low to Moderate Income (LMI), with the least change occurring in the South and Midwest (Mitchell and Franco n.d.). These communities also remain highly segregated, with the exception of those undergoing some measure of gentrification and therefore displacement of the historic residents.

The ongoing refusal to address and remediate the impacts of redlining, particularly in poor urban communities, is part of a broader governmental indifference toward the plight of the urban poor. The lack of an adequate governmental response following Hurricane Katrina stemmed, in part, from this same indifference (Dreier 2006). Leading up to 2005, cities were largely facing a similar set of challenges, including the negative impacts from highway policies, zoning and tax policies that promoted segregations, housing policies that negatively impacted the urban poor, and general disinvestment. Although there was a brief period of improvement in the 1990s, these improvements were all reversed after 2001 by a federal policy landscape that treated poverty as a character flaw and urban issues as entirely local problems (Dreier 2006). This prevailing attitude contributed to the framing of Katrina victims (in one of the ten most segregated cities of the United States) as thugs, or as individuals who refused to comply with evacuation orders, and the city as deserving of the disaster (Cigler 2007).

Although the preceding examples clearly illustrate the disparities in the levels of vulnerability between communities, as well as the long-standing impacts of discriminatory practices, they do not necessarily provide empirical evidence regarding the increased levels of hazard exposure among the poor and minorities. More recently, some empirical studies have provided evidence that the legacy of segregation includes the concentration of historically underserved populations in at-risk areas (Sloan and Fowler 2015). The recent National Flood Insurance Program (NFIP) affordability study, required under the 2014 Homeowner Flood Insurance Affordability Act, found that lower-income families are more likely to live in high-risk flood zones in the majority of states (Grueskin 2018). A study in the Houston area found that poor neighborhoods tended to have lower elevations, along with an association between lower elevations and high concentrations of minorities and noncitizens. This same pattern is also found across the largest metropolitan areas in the United States, indicating that structural factors are involved (Lu 2017).

Focusing specifically on New Orleans, Zakour and Grogg (2018) found a history of concentrating black populations into neighborhoods with nuisances such as swamps, open sewers, and a lack of public services. These same neighborhoods were also the ones most adversely impacted by the construction of canals designed to protect more affluent, and white, areas. Although flooding from Hurricane Katrina impacted many neighborhoods, 60% of the residences inhabited by black families were flooded for over ten days, as opposed to only 24% of those inhabited by white families (Zakour and Grogg 2018). This is consistent with Campanella's assertion (2018) that African American populations were more concentrated in the eastern portion of the Parish, in areas at greater risk of hurricane surge due to lower elevations, the degradation of protective wetlands, and the creation of man-made navigation canals—all leading to a greater number of African American victims in

Hurricane Katrina. These patterns of segregation and uneven distribution of risk continued following Hurricane Katrina when both white and black populations shifted out of the flooded areas (reducing the population of those areas by 37%) while the Hispanic population in the flooded zone increased by 10% (Campanella 2018).

In addition to the patterns of population distribution, environmental degradation, and land use that served to place minority population in at-risk places, the prioritization of corporate profits over the health and well-being of people of color has led to the concentration of dangerous environmental activities in and near minority communities (Maxwell 2018). The environmental justice literature has clearly documented the racial and socio-economic disparities in the siting of hazardous or polluting facilities (Zhang 2010). One example of this is cancer alley in Louisiana, an 85-mile stretch of land containing over 150 refineries and industrial plants, and populated primarily by people of color (Maxwell 2018). Even when making residential housing decisions, such as the purchase of a home, minority and low-income households have been found to be the most poorly informed regarding local hazards and toxic risks (Zhang 2010).

6.2 The Safety Premium

Disaster risk reduction is often presented as a local, or even personal, responsibility in the face of increasing disaster frequency and losses. This framing of risk reduction ignores the unequal creation of disaster risk and the various factors that impact the ability of certain groups to pay a safety premium. Climate adaptation and disaster risk reduction policies often ignore the very real costs of such measures, including not just financial burdens but also impacts to individual identity and social capital from changing communities or forced relocation. Climate adaptation can cause or exacerbate trauma, particularly in the face of a loss of individual agency, or for those communities of place.

Efforts at increasing safety through land-use policies, building codes, and zoning can result in forced population migration, as those unable to pay the added costs of safety can no longer afford to live in the environment. In some communities, this process is occurring through climate gentrification, as described in earlier chapters. One example of this phenomenon can be found in recent mobile home trends. The increase in demand for higher-quality, better-built, and luxury models by more affluent retirees, coupled with higher standards in places such as the Florida Keys have created situations in which families who lose a mobile home to a hurricane cannot afford to replace it. Prices for new double-wide trailers, for example, have risen by over 20% since 2012 (Gopal 2017). As a result, families on fixed incomes who had previously utilized trailers as a lower-cost housing alternative find themselves competing in an overpriced rental market or simply having to relocate.

As disaster recovery processes force lower-income residents to abandon at-risk areas, real estate speculators are prepared to quickly purchase properties at steep discounts. In fact, these speculators often target low-income neighborhoods in recently damaged areas. This type of profiteering takes advantage of the inability of lower-income homeowners to cover upfront costs for repairs or to navigate the lengthy bureaucratic process of recovery. This type of speculative buying has been occurring in the neighborhoods impacted by Hurricane Harvey, with speculators quickly reselling properties for a profit without fully disclosing risks (Romero 2018). This lack of disclosure is also prevalent in the rental housing market (Chapman-Henderson and Rierson 2017).

Another way in which more affluent communities are better able to secure disaster risk reduction measures is through their increased ability to pursue legal actions following significant disaster impacts. Pursuing legal action requires time, resources, and access to expertise. A lawsuit is currently pending regarding the flooding of West Houston, as a result of releases from the Addicks and Barker dams. Although the impacted properties were located in the identified flood pool, the releases began earlier than expected, in the middle of the night with little or no notice to residents (Sims 2017). This lack of notice increased the impacts from the flooding and added to the trauma of the impacted homeowners. The lawsuit is particularly noteworthy because it may impact the ability of the United States Army Corps of Engineers to undertake controlled releases, in some cases increasing the risk to communities downstream. This kind of action, taken in a lower-income, or minority, neighborhood, might garner a very different response, including far less attention from attorneys. An additional factor in this case is the impact to corporations that had headquarters within the impacted areas. In fact, other subdivisions located within the reservoirs are not pursuing similar legal actions (Kimmelman 2017).

6.3 Considering Justice in Resilience

At the present time, the costs of climate adaptation are primarily born at the local or individual level, with consideration of fair and equal distribution. As described above, these costs include impacts to individual agency and well-being, along with financial impacts. Although some federal grant programs do provide limited funding for hazard mitigation and other measures, these programs are designed to benefit homeowners and require some competition based upon a benefit-cost calculation. These types of calculations privilege high-value properties that can show a history of high-value losses, above other impacts to communities or lower-value structures. Additionally, the majority of these grant programs require some personal financial investment through the use of matching funds or other mechanisms.[2]

In a more extreme, but timely, example, some communities in places such as Louisiana and Alaska have been facing the threat of relocation for several years, as climate change, land loss, permafrost melting, and other factors such as environmental degradation from man-made canals have exacerbated their risk. One example of a community currently considering a relocation project is Princeville, North Carolina, which initially began discussing buy-outs in the late 1990s. Princeville is the oldest town in the United States to have been chartered by freed slaves and was, as a result, located in otherwise undesirable flood-prone land. Flooding from the Tar River, as a result of hurricanes, has caused repetitive flooding and led to a renewed community effort at tackling relocation (Bidgood 2016). Government failures to invest in resilience and preparedness for communities of color, already situated in riskier locations, have led to an urgent need for adaptation without sufficient resources and without an effort to ensure just outcomes (Maxwell 2018).

Returning to the four principles that form the foundation for Just Recovery, current disaster risk reduction policies do not provide equal access to resources and programs. Furthermore, the community's transformative and adaptive capacity is not fully harnessed in support of a resilient recovery.

Notes

1. Although this chapter does not trace the history of discrimination and injustice back to colonial time, the political economy of vulnerability as it has impacted indigenous populations can be traced back to colonial governments. Colonial governments removed individual agency and community efficacy through the forced limitation of population mobility, imposition of economic systems, and forced settlement patterns (Oliver-Smith 1996).
2. Even when matching requirements are waived, grant recipients are still expected to cover upfront costs and to wait long periods of time for reimbursements.

References

Bidgood, J. (2016, December 9). A Wrenching Decision Where Black History and Flood Intertwine. *The New York Times*. Retrieved from https://nyti.ms/2hw362f.

Campanella, R. (2018). Settlement Shifts in the Wake of Catastrophe. In M. J. Zakour, N. B. Mock, & P. Kadetz (Eds.), *Creating Katrina, Rebuilding Resilience: Lessons from New Orleans on Vulnerability and Resiliency* (pp. 25–44). Oxford: Butterworth–Heinemann.

Chapman-Henderson, L., & Rierson, A. K. (2017). Learning from the 2017 Disasters to Create a Reliably Resilient U.S. Federal Alliance for Safe Homes.

Cigler, B. A. (2007). The 'Big Question' of Katrina and the 2005 Great Flood of New Orleans. *Public Administration Review*, Special Issue: 64–77.

Dreier, P. (2006, March). Katrina and Power in America. *Urban Affairs Review*, 41(4), 528–549.

Gopal, P. (2017, November 21). Mobile Homes Are So Expensive Now, Hurricane Victims Can't Afford Them. *Bloomberg Businessweek*.

Grueskin, C. (2018, April). Study Finds Flood Insurance Affordability an Issue for Poor in Louisiana and Nationally. *The Advocate*.

Hardy, R. D., Milligan, R. A., & Heynen, N. (2017). Racial Coastal Formation: The Environmental Justice of Colorblind Adaption Planning for Sea-Level Rise. *Geoforum, 87*, 62–72.

Kimmelman, M. (2017, November 11). Lessons from Hurricane Harvey: Houston's Struggle Is America's Tale. Retreived from: https://www.nytimes.com/interactive/2017/11/11/climate/houston-flooding-climate.html

Lu, Y. (2017). Hurricane Flooding and Environmental Inequality: Do Disadvantaged Neighborhoods Have Lower Elevations? *Socius: Sociological Research for a Dynamic World, 3*, 1–3.

Maxwell, C. (2018, April 5). *America's Sordid Legacy on Race and Disaster Recovery*. Center for American Progress.

Mitchel, B., & Franco, J. (n.d.). *Holc 'Redlining' Maps: The Persistent Structure of Segregation and Economic Inequality*. NCRC Research.

NRC. (2012).

Oliver-Smith, A. (1996). Anthropological Research on Hazards and Disasters. *Annual Review of Anthropology, 25*(1), 303–328.

Romero, S. (2018, March 23). Houston Speculators Make a Fast Buck from Storm's Misery. *The New York Times*. Retrieved from https://nyti.ms/2g6scex.

Sims, S. (2017, November 16). The U.S. Flooded One of Houston's Richest Neighborhoods to Save Everyone. *Bloomberg Businessweek*.

Sloan, M., & Fowler, D. (2015). Lessons from Texas: 10 Years of Disaster Recovery Examined. *Texas Appleseed*. Retrieved from https://www.texasappleseed.org/.

United States, Government Accountability Office. (2014, December). *FEMA Has Improved Disaster Aid Verification but Could Act to Further Limit Improper Assistance*. National Research Council. Affordability of National Flood Insurance Program Premiums—Report 1.

Zakour, M. J., & Grogg, K. (2018). Three Centuries in the Making: Hurricane Katrina from an Historical Perspective. In M. J. Zakour, N. B. Mock, & P. Kadetz (Eds.), *Creating Katrina, Rebuilding Resilience: Lessons from New Orleans on Vulnerability and Resiliency* (pp. 159–192). Oxford: Butterworth–Heinemann.

Zhang, Y. (2010). Residential Housing Choice in a Multihazard Environment: Implications for Natural Hazards Mitigation and Community Environmental Justice. *Journal of Planning Education and Research, 30*(2), 117–131.

CHAPTER 7

Disparate Outcomes

Abstract This chapter takes a closer look at a series of recent disasters in order to highlight the disparate outcomes that resulted from the trends and covered in the preceding chapters. These disaster events include Superstorm Sandy, Hurricane Maria, and Hurricane Harvey. Previous chapters provided multiple examples from Superstorm Sandy and Hurricane Katrina, but a more in-depth discussion of disparate outcomes is provided in this chapter. An analysis of available texts, narratives, and reports, coupled with academic literature, forms the basis for the contents of the chapter.

These examples align with the Vulnerability-Plus model through the intersection of the risk creation described in the preceding chapter with the socio-political framing of undeserving victims and complex policy systems coupled with implementation failures, all leading to reductions in collective efficacy and individual agency in the face of, and following, natural events such as Hurricane Katrina. Political economic restructuring, using the disaster as a catalyst, increases vulnerability, creating a dynamic cycle through the reduction of access to capitals and limitations in the exercise of individual agency. The resulting disparate outcomes include increases in poverty and economic hardship, the loss of public housing, unique challenges for renters, and displacement.

Keywords Neoliberalism • Disaster capitalism • Disaster recovery • Vulnerability Plus • Disaster injustice

This chapter looks at a series of recent disasters in order to highlight the disparate outcomes that resulted from the various factors covered in the preceding chapters. These disaster events include Superstorm Sandy, Hurricane Maria, and Hurricane Harvey. Previous chapters have provided multiple examples from Superstorm Sandy and Hurricane Katrina, but a more in-depth discussion of these outcomes is merited. An analysis of available texts, narratives, and reports, coupled with academic literature, forms the basis for the contents of this chapter.

Preliminary analyses of the impacts of Hurricane Maria indicate that the adverse impacts in Puerto Rico have disproportionately impacted the most vulnerable communities and households, primarily those least able to access relief due to mobility or geographic isolation (Sou and Aponte-Gonzalez 2017). Similar impacts are being reported in Texas, following Hurricane Harvey, where immigrant families are twice as likely to have a loss of employment or income, less likely to receive disaster assistance, and less likely to have insurance (Reed 2018). The prevalence of ditch drainage in minority neighborhoods impacted by Hurricane Harvey, due to a historic lack of investment in infrastructure, also exacerbated flooding.

These examples align with the Vulnerability-Plus model through the intersection of the risk creation described in the preceding chapter with the socio-political framing of undeserving victims and complex policy systems coupled with implementation failures, all leading to reductions in collective efficacy and individual agency in the face of, and following, natural events such as Hurricane Katrina. Political economic restructuring, using the disaster as a catalyst, increases vulnerability, creating a dynamic cycle through the reduction of access to capitals and limitations in the exercise of individual agency (Zakour et al. 2018). The resulting disparate outcomes include increases in poverty and economic hardship, the loss of public housing, unique challenges for renters, and displacement.

7.1 Poverty and Economic Hardship

Zakour et al. (2018, p. 6) attest that Hurricane Katrina perpetuated and created a cycle of impoverishment due to "environmental injustice, including the inequitable accessibility to pre-Katrina disaster preparedness and post-Katrina recovery resources; damage to social infrastructure, including health and human service provisions; and, a markedly neoliberal approach to recovery aid." This combination of pressures increased negative economic and social outcomes, not just for families already below the poverty line but also for those precariously positioned within the lower middle class. Detrimental impacts to a struggling middle class have been clearly documented on St. Thomas, St. John, and St. Croix following Hurricane Maria (Craig 2018). Inequitable access to post-event recovery resources can also be seen on the US Virgin Islands as the poor face many more hurdles to recover than the much wealthier second homeowners, many of whom have made repairs.

The assertion that socio-economic conditions resulted in differential access to recovery resources following Hurricane Katrina was also made by Kamel (2012) in a study of repopulation rates and recovery patterns which found that the areas with the highest concentration of lower-income households, renters, and minorities experienced the greatest amount of population loss. Given the data from the previous chapter indicating higher concentrations of vulnerable populations within more at-risk areas, this finding can be considered to be at least partially connected to a greater level of damages. However, the study also found that access to federal recovery resources was not proportionate to the actual damage levels. This, of course, echoes what has already been highlighted in Chap. 4—Public Policy and Legislation, namely, that current policies are not designed or implemented to assist those most in need; simply stated, the approach is technocratic at best and not informed by any goals of fair or just distribution.

One of the most often touted mechanisms for improved recovery outcomes, flood insurance, has been shown repeatedly to be adopted at much lower rates among lower-income families. This trend can only be expected to continue as the price of insurance rises in the face of increasing disaster risk and losses. Lee (2018) argues that not only are low-income neighborhoods impacted most adversely by disasters but also these effects are long-lasting, perpetuating cycles of risk and poverty. One way in which risk is perpetuated, and increased, is through the habitation of increasingly damaged housing stock following subsequent events. The poor are more likely to be unable to adequately maintain their homes, resulting in impacts to recovery assistance from deferred maintenance, which, coupled with the lack of a financial safety net, can lead to improper repairs (Norton et al. 2018). These two factors, the lack of flood insurance and the habitation of damaged housing stock, further serve to limit the poor's access to recovery resources and illustrate Zakour's statement that access to preparedness resources is also inequitable. In Puerto Rico, for example, only around 2% of homeowners carried flood insurance (Oxfam 2018). As might be expected, those with less access to financial capital are less able to make repairs, further perpetuating the decline in housing stock and reducing the value of properties, a significant asset for the poor (Lee 2018). The inability to fully prepare and to update or maintain homes, particularly in the face of the structural factors outlined in the preceding narrative, is then taken as proof of undeservedness, as an unwillingness to engage in self-protective behaviors.

The perpetuation of a cycle of poverty following a natural disaster is clearly evident in Puerto Rico, with the poverty rate rising from 44–52% following Hurricane Maria. Despite undertaking self-recovery actions and turning to social capital networks for support, over 1 million households require further assistance in the face of rising costs for household goods and foods, coupled with limitations to income earning activities stemming from damages to homes and places of employment (Sou and Aponte-Gonzalez 2017). In some cases, this cycle of poverty is further perpetuated by the exodus of middle-class and wealthy residents, leaving just the poor who cannot afford the cost of leaving and a reduced tax base struggling to sustain government services (Flavelle 2018). These change processes within neighborhoods and communities, whether the cycle of attrition and decline described above or climate-driven gentrification in areas seen as more desirable, are accelerated by natural disasters due to the influx of resources and the direct damages. Increased poverty rates often result from the reduction of property values and population movement (Lee 2018).

Another way through which low and even moderate-income households are found to have disproportionate access to recovery resources is through the lack of equitable access to the benefits from rebuilding projects in the areas of housing and infrastructure (Sloan and Fowler 2015). These projects often produce benefits outside the areas of greatest impact, serving to promote a pre-existing economic development agenda at the expense of recovery for the poor, as was seen in Gulfport where Community Development Block Grant Disaster Recovery Assistance Program (CDBG-DR) funds following Hurricane Katrina were utilized for a port expansion project.

7.2 Impacts to Public Housing

Historically, catastrophic disasters have been utilized to further policies geared at the replacement of public housing stock with mixed-income developments, in most cases, displacing large number of families. Although there are certainly arguments that mixed-income housing models provide some benefits to the low-income residents of the developments, the number of units available has consistently been drastically reduced, and public housing is often not replaced once it has been damaged (Flavelle 2018). For example, following Hurricane Ike, only 40 of 500 public housing units were rebuilt (Kelly et al. 2017). Similarly, New Orleans lost over half of its public housing residences following Hurricane Katrina, including units with minimal damages where residents were not even allowed to enter and obtain their belongings. In fact, of the four neighborhoods which contained less than half of their overall pre-Katrina population in 2017, three had been public housing sites demolished to create mixed-income housing at a lower density (Data Center 2017). Several proposals to rebuild affordable housing in Mississippi and Louisiana were rejected by communities and city governments, with low-income housing often excluded from CDBG-DR Action Plans (Scurfield 2008). The failure to ensure that fair housing and civil rights regulations are enforced during housing recovery further exacerbates this challenge (Sloan and Fowler 2015). Hurricane Harvey, which damaged over 200,000 subsidized rental units, will prove to be another test of socio-political willingness to promote justice in housing (Kelly et al. 2017)

7.3 Renters

As has been previously stated in this text, renters are typically disproportionately affected by natural disasters and then by the federal recovery policies that fail to meet their needs. The challenges faced by renters include not only damages to current rental housing stock but also changes in affordability following an event and limitations in their ability to prepare, creating a similar dynamic to that discussed in the prior section where limited access to preparedness activities results in both increased vulnerability and the perception of the victims as unwilling to play a role in their own safety and, therefore, undeserving of assistance. Following Hurricane Katrina, the recovery of renters lagged behind that of the general population with a lack of affordable housing available for low-income families (Scurfield 2008; Kamel 2012).

Research conducted in California and Louisiana found several barriers in place, preventing, or limiting, the barriers of tenants to preparedness in the face of joint natural and technological disasters. These barriers include a lack of resources, disincentives for mitigation, which might increase the value of the property (which is also not their own) and therefore raise rents, and the high turnover, which can limit exposure to local public safety and preparedness messaging (Burby 2006).

In addition to the direct loss of rental housing as a result of high winds, flooding, or other hazards, the price of rental housing often increases following a disaster, as is predicted to near Houston (Kelly et al. 2017). There are several reasons for this price increase, including the shortage of rental units even prior to the loss of further units to the event, and the influx of families, including middle-class homeowners, seeking temporary rentals during rebuilding.

7.4 Displacement

In addition to the displacement of the poor, including those in public housing and renters, additional displacement can occur as a result of the added costs of returning, the safety premium described in the previous chapter, as well as due to the slow pace of recovery and a lack of basic services, as is currently been seen in Puerto Rico. Following Hurricane Maria, almost 400,000 individuals left Puerto Rico for the mainland, with the majority (around 150,000) going to Florida (Melgar 2018). Puerto Rican families left the island due to factors such as delays in repairing the power grid,[1] lack of access to water, failure of health infrastructure, damages to schools,[2] and loss of employment. The situation in Puerto Rico has been described by Oxfam (2018) as a humanitarian crisis, where the federal government's delays and inaction have impacted the most vulnerable, prioritizing process over people, in an example of administrative evil. Around 472,000 housing units on the island were seriously affected, of which over 80,000 were completely destroyed and the remainder experienced major damage (Oxfam 2018).

The Puerto Rican diaspora on the mainland has been pressured to return to Puerto Rico, through policies such as the refusal by Federal Emergency Management Agency (FEMA) to implement Direct Housing Assistance on the mainland, allowing for longer-term rental assistance and not just utilizing the Transitional Shelter Assistance (TSA) program, which is only designed for temporary assistance with hotel and motel-based lodging. Puerto Ricans depending upon TSA have faced repeated threats of voucher loss, including multiple deadlines for the program, leaving families wondering for several days if they will be homeless (Hernandez 2018). In each case, advocates such as Latino Justice have scrambled to file suits preventing the end of the program, often securing limited extensions. However, many Puerto Rican families have had to endure extended housing insecurity, exacerbating the trauma of displacement from their homes.

Displacement is also occurring in parts of the Florida Keys following Hurricane Irma, where the strong building code is effectively prohibiting residents from replacing or substantially repairing damaged mobile homes (Flavelle 2018). Although the stronger building codes do serve a public safety and disaster risk reduction function, they also create a situation in which existing residents are displaced in favor of those who can afford to meet the higher standard. Hurricane Irma damaged or destroyed around 1000 trailers and mobile homes across the impacted areas. The replacement of the trailers with costlier, and safer, homes constitutes a form of climate gentrification, leaving residents no choice but to either cover the cost of more expensive accommodations or leave. Increased costs of rebuilding stem not only from the higher construction costs, but also from increasing insurance premiums, and in some cases local taxes as bonds are passed for flood control efforts.

7.5 JUST RECOVERY

Returning to the principle of Just Recovery, one based upon capacities justice and the concept of well-being, it is critical to acknowledge that neither distributive nor participatory goals are being met by current recovery strategies. The comparison between the efforts in Texas and Puerto Rico alone presents a pressing example of the failure to fairly distribute resources when disparities in political power are present and when populations are judged undeserving. The following statistics, taken from Vinik (2018), illustrate the stark difference[3] between the two responses[4]:

- Deployment of Helicopters: 73 were deployed to Houston within six days of Harvey, whereas only 70 were deployed to Puerto Rico over the much longer time span of three weeks.
- Deployment of Personnel: **30,000** personnel were in Houston within nine days, compared to **10,000** in Puerto Rico within a similar time period.[5]
- Approval of Permanent Disaster Work: FEMA approved permanent disaster work for Texas within 10 days, as opposed to a delay of 43 days for Puerto Rico.

- Feeding Operations: 5.1 million meals and 4.5 million liters of water were deployed to Houston within nine days, whereas only 1.6 million meals and 2.8 million liters of water were deployed to Puerto Rico within a similar time period.
- Individual Assistance (IA): $141.8 million in IA was approved in the nine days following Harvey, as opposed to only $6.2 million in IA[6] following Maria.

Another comparison can be drawn simply through looking at power grid recovery following hurricanes. Some differences are due to the failure of Puerto Rico Electric Power Authority (PREPA) to immediately activate mutual aid agreements, as well as to the lack of a common language and terminology once those agreements were put to use. However, PREPA also lacked a command structure, and general capacity, due in part to a history of financial hardship (Ferris 2018). The following grid recovery comparisons are taken from Ferris (2018):

- 3000 miles of transmission lines and 263 substations were damaged by Hurricane Katrina, leaving over 1 million customers without power; three-fourth were restored within two weeks.
- 8.6 million customers were without power following Superstorm Sandy; 99% were restored within two weeks.
- 6 million customers were left without power following Hurricane Irma, 95% were restored within two weeks.

The glaring differences in resource allocation and disparate recovery outcomes do not adhere to the principles of PPFPE. The stark difference in treatment of Puerto Rico had not been justified, despite some arguments made regarding the local challenges.

Notes

1. Puerto Rico's blackout is the second largest on record, measured in customer hours, with over 60,000 families remaining without power in April of 2018 (Irfan 2018). Hurricane Maria knocked out 80% of the power lines across the island, with an estimated repair bill of over $17.6 billion.
2. Over 70 schools were severely damaged or destroyed (Oxfam 2018).

3. It is important to acknowledge that Puerto Rico's geography, and geographic location, did serve to make some elements of response more challenging than they might be on the mainland.
4. The delay in responding to a humanitarian crisis in Puerto Rico stands in sharp contrast to the speed with which protestors were met with tear gas and pepper spray following the release of a fiscal plan for Puerto Rico in April of 2018 that included austerity measures (Hernandez 2018).
5. Additionally, more experienced and senior personnel were sent to Houston, despite the scale of the challenges and need in Puerto Rico.
6. Applicants for assistance in Puerto Rico faced extraordinarily high rates of denials, with two of every five considered ineligible, and around 80% of subsequent appeals declined (Hernandez 2018). These denials were due to problems with the documentation of title, social security numbers, damage assessments, and other factors.

References

Burby, R. J. (2006). Hurricane Katrina and the Paradoxes of Government Disaster Policy: Bringing About Wise Governmental Decision for Hazardous Areas. *Annals of the American Academy of Political and Social Science, 604*, 171–191.

Craig, T. (2018, June). Virgin Islands Hurricane Recovery Efforts Hobbled by Cash Shortage. *The Washington Post.*

Ferris, D. (2018, April 19). How Puerto Rico Became the Worst Grid Disaster. *E and E News.*

Flavelle, C. (2018, November 8). Hurricanes Highlight Failure to Enforce Flood Insurance Rules. *Bloomberg.*

Hernandez, A. R. (2018, April 28). Sluggish Recovery from Hurricane Maria Reignites Calls for Puerto Rico's Statehood, Independence. *The Washington Post.*

Irfan, U. (2018, April 15). Puerto Rico's Blackout Is Now the Second Largest on Record Worldwide. Retrieved from https://www.vox.com/2018/4/13/17229172/puerto-rico-blackout-hurricane-maria.

Kamel, N. (2012). Social Marginalisation, Federal Assistance and Repopulation Patterns in the New Orleans Metropolitan Area Following Hurricane Katrina. *Urban Studies, 49*(14), 3211–3231.

Kelly, C., Costa, K., & Edelman, S. (2017, October 3). *Safe, Strong, and Just Rebuilding After Hurricane Harvey, Irma, and Maria Recommendations.* Center for American Progress.

Lee, D. (2018). The Impact of Natural Disasters on Neighborhood Poverty Rate: A Neighborhood Change Perspective. *Journal of Planning Education and Research, 00*(0), 1–13.

Melgar, M. E. L. (2018, May 11). Mapping Puerto Rico's Hurricane Mitigation with Mobile Phone Data. *City Lab.*

Norton, R., MacClune, K., Venkateswaran, K., & Szönyi, M. (2018). *Houston and Hurricane Harvey: A Call to Action.* Zurich, Switzerland: Zurich Insurance Company Ltd.

Oxfam Research Report. (2018). *Far from Recovery: Puerto Rico Six Months After Hurricane Maria.*

Reed, E. (2018, April 30). Immigrant Families Have Not Bounced Back from Hurricane Harvey. *The Street.*

Scurfield, R. (2008). Post-Katrina Storm Disorder and Recovery in Mississippi More Than 2 Years Later. *Traumatology, 14*(2), 88–106.

Sloan, M., & Fowler, D. (2015). Lessons from Texas: 10 Years of Disaster Recovery Examined. *Texas Appleseed.*

Sou, G., & Aponte-Gonzalez, F. (2017, December). *Making Efforts Count After Irma and Maria: Household Relief and Recovery in Puerto Rico.* University of Manchester, UK: Policy Brief.

Vinik, D. (May 29, 2018). 'People Just Give Up': Low-Income Hurricane Victims Slam Federal Relief Programs.

Zakour, M. J., Mock, N. B., & Kadetz, P. (2018). Editor's Introduction: The Voices of the Barefoot Scholars. In M. J. Zakour, N. B. Mock, & P. Kadetz (Eds.), *Creating Katrina, Rebuilding Resilience: Lessons from New Orleans on Vulnerability and Resiliency* (pp. 3–23). Oxford: Butterworth-Heinemann.

CHAPTER 8

Conclusion: Resilience for Whom?

Abstract This chapter constitutes the final call to action, asking policy makers, emergency managers, disaster professionals, and other interested parties to take a closer look at the role that current policies around disasters play in creating and perpetuating injustice. As the previous chapters bring to light, disaster recovery policies and programs have routinely and repeatedly failed to prioritize human rights and failed to acknowledge the dynamic pressures and complex history of disaster risk creation in the United States. The prevailing narratives surrounding climate adaptation and disaster risk reduction, both extremely pressing in the current fiscal and environmental climate, have been willfully colorblind, perpetuating existing inequalities while fostering a climate of blame. The chapter argues that the definition and framing of resilience utilized by policy makers, professionals, and communities cannot be value neutral; the question must always be asked: resilience for whom?

Keywords Capacities justice • Human rights • Disaster recovery • Resilience

This book constitutes a call to action, asking policy makers, emergency managers, disaster professionals, and other interested parties to take a closer look at the role that current policies around disasters play in creating and perpetuating injustice. This call to action demands that we acknowledge the fact that, even when well-intentioned, disaster recovery policies and programs have routinely and repeatedly failed to prioritize human rights and failed to acknowledge the dynamic pressures and complex history of disaster risk creation in the United States. To take it one step further, one might say that the prevailing narratives surrounding climate adaptation and disaster risk reduction, both extremely pressing in the current fiscal and environmental climate, have been willfully colorblind, perpetuating existing inequalities while fostering a climate of blame. At the minimum, these narratives have focused on technical elements of risk and not on justice. They have focused on monetary damages, and on repairing damages, but not on making people whole.

The call of action is this: The definition and framing of resilience utilized by policy makers, professionals, and communities cannot be value neutral; the question must always be asked: resilience for whom? When resilience is defined by broad-scale metrics that ignore individual experiences and power structures, places may simply become more resilient through population displacement, further concentrating vulnerability among those groups unable to afford the safety premium and treated as undeserving victims by the same systems that exacerbate the vulnerability and guarantee disparate outcomes. Displacement of the vulnerable is not resilience. Unfortunately, the majority of the existing narratives around resilience[1] tend to ignore history and context, perpetuating systemic injustice and offering resilience only to a privileged class. Even when access to disaster risk reduction is offered, in terms of the ability to apply for programs, the conditions of the programs render participation impossible for many. Equitable and Just Recovery, as recommended by the Center for American Progress, must prioritize communities with the fewest resources and greatest needs (Kelly et al. 2017). It must focus resources where infrastructure was previously failing, homes lacked clear title, and residents were unable to afford maintenance. Just Recovery must be more than a return to the status quo.

8.1 Revisiting Just Recovery

This text proposes a series of principles underlying the concept of Just Recovery. Although many researchers and policy makers have acknowledged the existence of inequalities and unjust outcomes, justice paradigms have not informed policy making in any concrete way. The four principles are intended to launch a conversation, serving as a bridge from theory to practice. If these principles are taken into account in policy making, planning, and implementation, more just outcomes will be the result.

Principle #1: Just Recovery requires that all community members (regardless of their socio-economic status, race, gender, sexual identification, land tenure, etc.) be provided with the ability to exercise their agency fully through free and informed choice in support of their personal well-being. Full and informed choice is not possible if there is any coercion, exclusion from public policies, or other barriers to full participation. Furthermore, agency cannot be fully exercised if a full and complete array of options is not understood and made available within a timely fashion and in means that are accessible.

Principle #2: Just Recovery begins from the principle of prima facie political equality (PPFPE), which clearly establishes that any different or unequal treatment must be justified by the discriminator; only equality is inherently defensible. Any expectation that disaster victims or communities prove their deservedness in the face of impartial bureaucratic processes puts the onus on the victims to justify the need for equal treatment and fails the PPFPE test.

Principle #3: Just Recovery requires the full harnessing of the communities' transformative and adaptive capacity, honoring their definitions of resilience, in order to reduce risks for the future. Holistic disaster risk reduction is not possible without acknowledging existing patterns of unequal distribution of risk. It is not sufficient to simply mitigate against some current risks in rebuilding; instead underlying structures and patterns must be questioned. Colorblind and a-historic recovery, which does not consider context, is not Just.

Principle #4: Just Recovery is not possible without equal access to resources and programs, including full participating in decision-making processes that govern resource allocation, future development, and other related functions.

Asking the question, resilience for whom, is imperative if we are to achieve disaster justice. These principles can serve to translate the concept of disaster justice into a set of guidelines that can serve to improve policies and their implementation. Other researchers and policy makers have made specific recommendations that are in line with these principles. For example, Kelly et al. (2017) specifically recommend that mortgage payments be deferred for 12 months during rebuilding and that some portion of insurance proceeds be permitted to be utilized for securing the property, allowing low-income homeowners to focus resources on recovery and preventing further damages as a result of delays in the assistance process.

Identifying concrete recommendations that support the four principles, along with the necessary changes to policies and regulations that allow for their implementation, is the next step. Therein lies the call to action.

8.2 Person and Community-Centered Recovery

Fundamentally, these principles and recommendations position persons and community at the center of recovery processes. They are informed by human rights principles, prioritizing needs above market value. Although they do not explicitly reference processes of public deliberation, full inclusion in these processes, such as in the creation of Action Plans, is required in order to meet the principles. As Sloan and Fowler (2015) observe, civil rights and fair housing requirements can be better utilized in order to achieve just outcomes. Returning to procedural, or participatory, justice, this requires the inclusion of all groups in the deliberative processes, consultations over time, access to information, including science, shared decision-making authority, and resources (Hunold and Young 1998).

A justice-based approach requires revisiting current policies and frameworks, including the mechanisms through which they are implemented, through a focus on distributive outcomes, well-being, and survivor agency. A process that prioritizes people over process requires flexibility and an awareness of the harms that are wrought by administrative evil. Bureaucratic policies and procedures, including the funding allocated for disaster risk reduction and recovery, reflect socio-political values and assumptions of deservedness. A far more just approach is one in which unmet needs drive resources, not local capacity or political considerations. Institutionalizing of Community Development Block Grant Disaster Recovery Assistance Program (CDBG-DR) rules and guidance, including more oversight and technical assistance, is one mechanism for ensuring greater consistency and incorporation of the principles described above (Sloan and Fowler 2015).

Note

1. This is particularly true when one looks at operational definitions utilized in practice and policy, as well as those definitions that form the basis of data-driven indices.

REFERENCES

Hunold, C., & Young, I. M. (1998). Justice, Democracy, and Hazardous Siting. *Political Studies, XLVI*, 82–95.

Kelly, C., Costa, K., & Edelman, S. (2017, October). *Safe, Strong, and Just Rebuilding After Hurricanes Harvey, Irma, and Maria: A Policy Road Map for Congress*. Center for American Progress.

Sloan, M., & Fowler, D. (2015). Lessons from Texas: 10 Years of Disaster Recovery Examined. *Texas Appleseed*.

Index[1]

A
Abbott, G. (Governor), 69
Access model, 10
Ackerman, B., 9
Action Plan, 51, 65n6, 76, 91, 100
Adaptive capacity, 9, 21, 52, 84, 99
Addicks dam, 83
Administrative evil, 47, 48, 53n1, 68, 71, 92, 100
Allen, Barbara C., 4, 8, 9, 17
Altruistic punishment, 32

B
Barker dam, 83
Benefit-cost approach, 16, 58, 63, 84
Bring New Orleans Back Commission, 17, 51
Bronze Star LLC, 39
Buffalo Creek Disaster, 29
Bureaucratic discretion, 48, 53n2, 68
Bush, George, 37

C
Campanella, R., 81
Capabilities approach, 8, 9
Capacities justice, 4, 46, 76, 93
Capital
 cognitive, 14, 15
 social, 10, 13–15, 20, 27, 28, 46, 49–51
 structural, 14, 15
CDBG-DR, *see* Community Development Block Grant Disaster Recovery Assistance Program
Center for American Progress, 98
Civil unrest, 47
Climate adaptation, 4, 18, 80, 82, 84, 98
Climate change, 2, 33, 84
Climate gentrification, 79, 80, 82, 90, 93
Coastal gentrification, 1
Coastal Zone Management Act, 7
Code of Federal Regulations, 72

[1]Note: Page numbers followed by 'n' refer to notes.

Cognitive capital, 14, 15
Collective action, 10, 14
Collective efficacy, 13–15, 49–51, 87, 88
  defined, 49
Community-centered recovery, 100
Community Development Block Grant Disaster Recovery Assistance Program (CDBG-DR), 51, 61, 70, 72, 74, 75, 90, 100
  Action Plan, 65n6, 76, 91
Community efficacy, 9, 33, 85n1
Community Rating System, 65n6
Community resilience, 63
  social capital, role of, 13
Constitution of the United States, 57
Corrosive community, 28, 29, 32–34, 46, 49
Cultural justice, 4, 8

D

Davis-Bacon Act, 37
Deepwater Horizon Oil Spill, 29, 72
Deferred maintenance, 62, 74, 89
Democracy free zone, 12
Department of Homeland Security (DHS), 37, 47
Department of Housing and Urban Development (HUD), 28
Deservedness, 12, 13, 20, 28, 32, 40, 41, 47, 58, 70, 90, 99, 100
Deserving corporations, 35–38
  Puerto Rico, 38–39
Deserving victims, 11–13, 27–41, 46, 48, 70
  corrosive community and, 32–34
  duplication of benefits and appeals, 30–32
  government corruption, 40
  and justice, 40–41
  Puerto Rico, 34–35

DHS, *see* Department of Homeland Security
Direct Housing Assistance, 92
Direct Lease program, 48
Disaster, 27, 29, 31, 36, 38, 40
  assistance, 56, 58, 59
  capitalism, 12
  justice, 10, 19
  market, 12
  policy, 56–57
  recovery, 2, 4, 9, 11–14, 16, 17, 19, 20, 27, 28, 30, 33, 35, 37, 45, 46, 48, 51, 56, 59, 62–64, 69, 88–92, 98
  resilience, 4, 9–11, 19–20
  risk creation, 80–82
  risk reduction, 18, 36, 58, 59, 63, 64, 71, 79–84
  vulnerability, 10–11, 22n8
Disaster Assistance Centers, 72
Disaster Mitigation Act of 2000 (DMA2K), 56, 57
Disaster Recovery Reform Act of 2018, 65n3
Disparate outcomes, 2, 4, 18–19, 68, 71, 88–94
  impacts to public housing, 90–91
  Just Recovery, 93–94
  person and community-centered recovery, 100
  poverty and economic hardship, 88–90
  renters, 91–92
Distribution of goods, 7, 8, 75
Distributive justice, 4, 7–9, 40, 61
DMA2K, *see* Disaster Mitigation Act of 2000
Documentation of ownership, 41n3
Duplication of benefits, 28, 30–32, 36, 62, 73
Dust Bowl, 2

E
Eccleston, Collette P., 12
Economic development, 10, 75, 79, 90
Economic equality, 9
Economic hardship, 88–90
Emergency management, 36, 57, 60, 63
Environmental degradation, 5, 82, 84
Environmental injustice, 5, 41n2, 88
Environmental justice, 4–7, 9, 17, 82
Equality, 11
  economic, 9
  political, 9
  principle of prima facie political, 9, 19, 20, 41
Equal opportunity, 9
Equitable Recovery, 98
Equity, 40, 50, 76
Exxon Valdez Oil Spill, 29

F
Fair Housing Act of 1968, 75
Fairness, 7, 11, 75
Federal disaster assistance, 34, 35, 56, 57
Federal Disaster Relief Act, 56
Federal Emergency Management Agency (FEMA), 12, 28, 31, 32, 34–36, 38, 39, 41n3, 47, 48, 60–62, 69, 73, 74, 92, 93
Federalism, 55, 57–60
FEMA, *see* Federal Emergency Management Agency
Fleming, David A., 15
Forced migration, 11, 21n1, 21n4, 82
Foreign Exchange Management Act, 29
Fowler, D., 100
Free consent, 9
Full participation, 7, 8, 20, 21, 50, 99

G
General Land Office, 69
Global Platform for Disaster Risk Reduction (2017), 63
Government corruption, 13, 40
Great Houston Flood of 1935, 59
Great Mississippi River Flood of 1927, 2
Grogg, K., 81

H
Hardy, R. Dean, 18
Hazard mitigation, 16, 18, 56–58, 60, 63, 79, 84
Hazard Mitigation Grant Program (HMGP), 60
HMGP, *see* Hazard Mitigation Grant Program
HOLC, *see* Home Owner's Loan Corporation
Holy Cross, 17
Home Owner's Loan Corporation (HOLC), 80
Homeowner Flood Insurance Affordability Act of 2014, 81
Homeowning nuclear families, 55
HUD, *see* Department of Housing and Urban Development
Human rights, 21n4, 63–64, 70
  institutional and structural violations of, 13
Hurricane Betsy, 32
Hurricane Dolly, 19
Hurricane Gustav, 19
Hurricane Harvey, 2, 3, 6, 19, 30, 49, 69, 70, 73, 74, 83, 88, 91, 93, 94
Hurricane Ike, 91
Hurricane Irma, 93
Hurricane Katrina, 2, 4, 6, 11, 12, 19, 28–31, 33, 37, 40, 46, 50, 52, 52n1, 53n3, 56, 61, 68, 69, 71, 72, 74, 75, 80–82, 88–91, 94

Hurricane Maria, 3, 28, 34–36, 38, 39, 48, 52n1, 61, 70, 88–90, 92, 94
Hyogo Framework, 56, 63, 64

I
IHP, *see* Individuals and Households Program
Implementation challenges, 17–18, 67–76
    current policies and regulations to promote just outcomes, using, 75
    current programs and policies, limitations of, 71–75
    existing policies and programs, 70–71
    and justice, 75–76
Individual agency, 45–51
Individual assistance, 60
Individuals and Households Program (IHP), 34, 60, 73
Informed consent, 9, 19–20
In-group vs. the out-group, 33
Institution of Occupational Safety and Health (IOSH), 37
Interim National Progress Report on the Implementation of the Hyogo Framework (2012), 64

J
Jerolleman, A., 56
Jim Crow laws, 64
Juan, San, 37
Justice, 11, 14, 18–20, 29, 38, 41n2, 48, 58, 63, 73, 91, 92, 99, 100
    capacities, 4, 46, 76, 93
    conceptualization of, 5–9
    cultural, 4, 8
    deserving victims and, 40–41
    disaster, 10, 19
    distributive, 4, 7–9, 40
    environmental, 4–7, 9, 17, 82
    implementation challenges and, 75–76
    litigation and, 64
    paradigm to disaster study, applying, 4–5
    post-disaster fraud and, 40–41
    procedural, 4, 7–9, 60
    public policy and, 64
    representative, 4, 8
    in resilience, 84
    social, 8, 9, 50
    survivor agency and, 52
Just Recovery, 5, 9–11, 19, 20, 21n6, 40, 41, 52, 62, 64, 72, 93–94, 98–100
    defined, 20–21

K
Kamel, Nabil, 73
Kelly, C., 99
Klein, Naomi, 12, 13
Klima, K., 56

L
Land-use decision, 57, 59
Laska, Shirley, 33, 48
Latino workers, 30, 41n5
Legislation, 16–17, 55–64
Lower Mississippi River Flood Control Act, 56

M
Maldistribution, 8
Marinov, Nikolay, 40
Mark, J., 47
Mental health problems, 50
Mississippi flood of 1927, 56

N
National Flood Insurance Program (NFIP), 56, 60, 62, 81
Natural hazards, 16, 29

Neoliberalism, 37
NFIP, *see* National Flood Insurance Program
NFIP Act, 56
Nikolova, Elena, 40
Non-disclosure, 7

O
Occupational Safety and Health Administration (OSHA), 37
Office of the Inspector General (OIG), 31, 37
OIG, *see* Office of the Inspector General
Oppression, 50, 51
Oxfam, 34, 92

P
PA, *see* Public Assistance
PAR, *see* Pressure and Release Model
Personal agency, 9
Person-centered recovery, 100
Pew Charitable Trusts, 60
Political equality, 9
Population shift, 1
Post-disaster fraud, 11–13, 27–41
  corrosive community and, 32–34
  duplication of benefits and appeals, 30–32
  government corruption, 40
  and justice, 40–41
Post-disaster victimization, 49
Poverty, 80, 81, 88–90
PPFPE, *see* Principle of prima facie political equality
Pre Disaster Mitigation Grant Program, 57
PREPA, *see* Puerto Rico Electric Power Authority
Pressure and Release Model (PAR), 10, 11

Principle of prima facie political equality (PPFPE), 9, 19, 20, 41, 94
Privatization, 36
Procedural justice, 4, 7–9, 60
  obstacles to, 8
Program implementation, 68–69
Project Impact, 65n5
Property rights, 8, 56
Public administration, 57
Public Administration Deductible, 58
Public Assistance (PA), 60, 61
Public Assistance Grant Program, 60
Public housing, 21n1, 48, 56, 71, 88, 92
  disparate outcomes, impacts of, 90–91
Public policy, 16–17, 55–64, 69, 99
Puerto Rico, 3, 29, 47, 61, 70, 88–90, 92–94, 94n1, 95n3–6
  Community Development Block Grant Disaster Recovery funds, 51
  deserving corporations, 38–39
  deserving victims, 34–35
Puerto Rico Electric Power Authority (PREPA), 37, 39, 94

R
Rapid innovation, 57
Rapidity, 10
Rawls, J., 9
Redundancy, 10
Rental stipends, 48
Renters, 91–92
Representative justice, 4, 8
Resilience, 40, 46, 50, 61, 63, 97–100
  justice in, 84
Road Home, 56, 62, 68, 72, 74, 75
Robert T. Stafford Disaster Relief and Emergency Assistance Act of 1988, 56, 61
Roberts, Patrick S., 48
Robustness, 10

S

Safety premium, 82–83
SBA, see Small Business Administration loans
Sendai Framework, 56, 63
SFHA, see Special Flood Hazard Area
Shrader-Frechette, Kristin, 9
Sloan, M., 100
Small Business Administration (SBA), 32, 60, 62, 73
Social capital, 10, 13–15, 20, 27, 28, 46, 49–51
Social injustice, 5, 33
Social justice, 8, 9, 50
Southern Poverty Law Center, 30
Special Flood Hazard Area (SFHA), 59
Spokane, Arnold R., 49
Stafford Act, 16, 56, 57
St. Bernard Parish, 52, 75
Street-level bureaucracy, 68
Structural capital, 14, 15
Structural violence, 11, 33, 46, 47, 51
Superstorm Sandy, 1, 16, 28, 31, 33, 56, 61, 68, 71, 88, 94
Survivor agency, 9, 13–14, 45–52
  collective efficacy and social capital, 49–51
  and justice, 52
  reclaiming, 51–52
Swager, Charles M., 10, 14, 21n8

T

Technological disaster, 28, 29
Texas General Land Office, 73
Texas Housers, 73
Texas Low Income Housing Information Service (Texas Housers), 19
Therapeutic community, 15, 28, 50
Toxic contamination, 28
Transformative capacity, 9
Transitional Shelter Assistance Program, 48
Tribute Contracting, 38–39

U

UNISDR, see United Nations Office for Disaster Risk Reduction
United Nations Office for Disaster Risk Reduction (UNISDR), 63
United States Army Corps of Engineers, 83
Universal Declaration of Human Rights, 64

V

Vinik, D., 74
Vulnerability, 29, 30, 33, 37, 58, 63
Vulnerability-Plus (V+) theory, 10, 11, 80, 88

W

Well-being, 2, 8, 20, 46, 52, 64, 75–76, 82, 84, 93
Wellness, 11
Whitefish Energy Holdings LLC, 37, 39
Whole Community Approach, 64
Wisner, B., 10

Z

Zakour, M. J., 9, 10, 14, 21n8, 81

Ingram Content Group UK Ltd.
Milton Keynes UK
UKHW020146050723
424579UK00004B/258